# SEX WARS

*How Hormones Drive Gender,
Race, and Culture Conflicts*

Roy Barzilai

## A Note on Cover Images

*The Fall of Man* (1508-1512)
Michelangelo Buonarroti Simoni (1475-1564)
Sistine Chapel. Paint on plaster.

The biblical story of the tree of knowledge beautifully depicts the prime human motives, associated with both body and mind: sex, temptation, authority, immortality, seeking knowledge and the morality of good and evil. Sex, life, and death of the body are contrasted with the mind acquiring moral reasoning and knowing the eternal truth of God.

The Christian doctrine of the fall is associated with the doctrine of Original Sin, the fall of the world into the sins of the flesh and sex. This pessimistic view of human nature is in contrast to Genesis 1 that depicts the creation of the world for the good of man and blesses him to "be fruitful and multiply." This dualism between creation and destruction, mind and body, and good and evil is the primary force in our religious and cultural worldviews, and as portrayed in *Sex Wars*, the course of human history can be framed as a pendulum continuously swinging between these two extremes.

# CONTENTS

*Preface*                                                                     *vii*

*Introduction*                                                                  *1*

1   ON HORMONES, FEELINGS, AND IDEAS: MIND–BODY
    DUALISM                                                                     21

2   DOMINANCE AND SUBMISSION                                                    51

3   CULTURAL EVOLUTION AND THE GENERATION GAP                                   83

4   THE SOCIAL ANIMAL                                                          105

5   THE ORIGINS OF SEX AND DEATH: SYMBIOGENESIS                                125

6   GENDER, RACE, AND CULTURE WARS                                             143

7   HOW SOLAR CYCLES DRIVE OUR EVOLUTION AND
    HISTORY                                                                    167

8   THE SUN, CLIMATE CHANGE, AND HUMAN CULTURE                                 199

9   HOW SEX HORMONES SHAPE OUR VIEW OF THE
    PHYSICAL UNIVERSE                                                          215

10  TRANSCENDING DETERMINISM                                                   241

*Notes*                                                                       *269*

# PREFACE

There are battles raging in our culture.

And whether we realize it or not, the defining theme of these battles is not simply what seems to propel clashes throughout history: religion or national borders or economics.

The defining theme is sex. These are the Sex Wars.

The sexual organization of society is at the center of the culture wars today, with the progressive wave of feminism and socialism in the West fundamentally transforming gender relations and our entire sociopolitical structure. The foundational institution of the family is disintegrating as marriage rates decline, divorce rates rise, and birth rates fall. We see an all-out attack on masculine values, coming mainly from the academic, leftist culture, along with a rise of homosexuality and even pansexuality with the notion of fluid gender identity prevailing in the movement toward transgenderism. At the same time, Islamic jihad threatens a return to primitivism and sexual repression, depression, and oppression in the subjugation of women as it wages its war of terrorism against the West. This encroachment into our culture brings with it a culture of destruction and death. The barbarians are storming the gates of the West as it is weakened, castrated, and rendered impotent from within and ready to be conquered and raped into submission from without.

The period of the ascendance of the West seems to be coming to an ugly end, losing its life energies in a similar manner that the aging human body fails, resulting in weakness, disease, fatigue, and depression, until the cycle of decay completes as the body finally succumbs to death. The life cycles of humans have been well documented by medical science; similarly the rise and fall of civilization has been documented by major historians, such as Edward Gibbons with his seminal writing, *The History of the Decline and Fall of the Roman Empire.* It is the goal of this book, *Sex Wars, to* integrate the two different disciplines, the humanities and biology, in order to further discover the mechanism that drives history to repeat itself according to biological, hormonal life cycles.

Building on my most recent book, *The Testosterone Hypothesis: How Hormones Regulate the Life Cycles of Civilization*, I will persevere in the quest to understand the biological forces that drive the recurring cycles of human expansion and rise, only to reach their peak and then fall back down again. In this work we will explore the connecting thread that explains what may seem to the unsuspecting eye a random stream of disconnected events. We shall see there actually is a single force, rooted in our biological makeup, our hormones, that appears in many different forms and drives the waves of social mood, shaping the evolution of our social life, the fluctuations of religious doctrines, cultural crusades, sexual norms, and even fashions trends. This force propels the very course of human history, regulating the recurring cycles in the advancement and decline of great civilizations.

As *Homo sapiens*, the rational man, we have a dual nature. The man of the mind is characterized by the faculty of reason, but humans are also social mammals, driven by animal instincts for survival, procreation, and social status. As individuals, sex is a defining feature of our personality and gender characteristics. As social animals it also shapes our social structure, the model for the organization of our society and playing an essential role in our biology as sexually reproducing organisms. This is evident in the culture wars taking place in our modern contemporary society. As we shall explore through this work, this dual nature of the human condition is at the core of the conflicts and the cultural storm we are experiencing today. Our biological drives function like a secret world, inhabiting the subconscious mind in the Freudian sense, working behind the scenes, forming our personal and social life in ways that we have not yet dared to imagine.

This fundamental transformation of our culture has led to a generation gap and conflict between the old, traditional values, social and political structures, and the new wave of change, led by the young face of the millennial generation, which calls for a different worldview to accommodate the new age. The values and traditions once cherished by the aging generation are challenged as arcane and even illegitimate by the rise of new generation with a different set of motivational drives.

Indicative of such differing motivations is the seemingly paradoxical phenomenon in which the feminists in the West are at war against Western men, but at the same time are incredibly welcoming to Third World, Islamic migrants—mostly men—who have no tradition of respect for women's rights and are commanded by their religious traditions to abuse women, beat them into submission, and deny them education and work opportunities. The influx of barbarian cultures, having no concept of liberty and individual rights—values that have been significant to the ascendancy of the West since the seventeenth century Enlightenment—means that Western civilization is in a rapid process of decline from within, as these demographic changes will eventually determine its future.

The feminist's multiculturalist ideology is powered by a socialist belief system, which holds that the West is led by power hungry, greedy, and egotistical men who built the exploitive capitalist system, and hence, is deemed repressive and evil toward women, minorities, and other cultures. To compensate for its past misgivings, this patriarchal culture must now humble itself and surrender its dominant position. From the feminist perspective, which dominates academia and forms the minds of the next generation, the demise of Western civilization is actually to be celebrated, given that this evil culture must be abandoned, conquered, and destroyed under the multicultural agenda. A new doctrine of "Original Sin" has been established by which Western men must recognize the evils of "toxic masculinity" and confess their sin of "white male privilege," thus surrendering their egotistical drive to pursue their own self-interest. The Western male and anyone complicit with his culture must submit and accept the new feminist-socialist regime along with the barbarian invasion of Islamic jihadists.

It cannot be stressed enough that through the universities and even earlier levels of education, students are actually being indoctrinated to disdain anything manly. According to this hate culture toward men, only when men are mentally and physically castrated and rendered impotent will their dominance and rape of women, minorities, and other societies finally come to its bitter end. Hence, as in the Marxist revolutionary ideology of the early twentieth century, the new form of cultural Marxism, or feminist takeover of society, will result in a feminist-socialist utopia in which men are obsolete. According to these radicals, every measure must be taken to bring this heavenly vision of society on earth. While this assertion might sound extreme, it will be discussed more thoroughly in chapter 2 when we address dominance and submission.

This political struggle manifests the spirit of our times, as our culture is being increasingly divided and split in an internal battle for its soul. We are witnessing a battle of the sexes that is rooted deep within the schizophrenic nature of the human brain, the masculine versus the feminine. This suggests that biological factors play a primary role in the development of our personality as the secret or "shadow" government that drives our gender and race politics, and hence divisive cultural and religious conflicts. These "wars" range from core issues such as feminism, sexual identification, and the definition of marriage, to our ideological and moral battles over the proper political system, whether for the free market, liberty, and capitalism or more socialism and greater government control of society and individuals.

On the national level in the United States, the presidential election of 2016 saw the gender war manifest itself most conspicuously with Hillary Clinton on the feminist side, as the Democratic socialist candidate, versus Donald Trump on the masculine side, as the Republican candidate for office. For the first time in the history of the nation, a woman was the candidate of a major party for the highest political office, the presidency of the United States. Hillary presented her femaleness as a primary virtue, her supporters believing that feminine values of empathy and compassion are superior to the crude and aggressive masculine behavior of her male opponent, Trump. One of the major controversies of the election was a moral and culture war on the issues of isolation and nationalism versus inclusiveness and globalism. Hillary called for the internationalist-socialist policy of open borders, allowing for foreign and Islamic immigration, and multiculturalism, while Trump stood strong for Western culture, Judeo-Christian values, and closed borders in a policy he framed as "America First." Donald Trump was elected president in the final results, but the country still remains largely divided. The moral, ideological, and political gulf between the two worldviews and political parties seems insurmountable.

Just as the gender wars cause a rise in divorce rates and a breakup of the marriage union, the US as whole is now experiencing a divorce between its feminine and masculine cultures that could lead to a breakup of its 240 year union. Sex wars may well bring about the fall of Western civilization as I predicted in my last book: Testosterone collapse will cause a profound culture clash and the subsequent demise of our civilization. This is evidenced by the likes of polls that reveal an increase in citizens who support their state's succession from the Republic. In California, for example, where twenty percent of those polled in 2014 wanted a "peaceful withdrawal from the union," thirty-three percent favored such a withdrawal when polled in 2017.[1]

After taking office in January 2017, President Trump quickly attempted to initiate a so-called "Muslim ban," a ban on people coming to the United States from seven Islamic countries that were identified by the Obama administration as posing an increased terror risk. Trump defended his executive order, saying, "This is not about religion, this is about terror and keeping our country safe."[2]

By contrast, throughout the administration of President Obama, an icon of the feminist–socialist left, he repeatedly defended Islam as a "religion of peace," while refusing to use the term *Islamic terror*, even when waves of mass terror attacks were being carried out by Islamic jihadists across the world, from the Middle East to Africa and in the West itself.

This proposed ban is indicative of the alpha-male dominance wars and the territorial imperative: the dominant, Western alpha male taking control of his territory—a response to the Islamic jihad of aggression and terror against the West, including literal rape. Speaking at Al-Aqsa Mosque in Jerusalem, Sheikh Muhammad Ayed encouraged Muslim migrants to "breed children" with Europeans to "conquer their countries." "We will trample them underfoot, Allah willing," the Imam vowed in response to his assertion that the West is only accepting refugees to increase their labor force.[3]

In defiance of the Trump administration's policies and exhibition of their vile hatred of Western men and traditional American culture, the feminists organized a nation-wide women's march the week after the president's inauguration, ironically aligning themselves with Islamic women who support Sharia law, which subjugates women to male oppression:

> Linda Sarsour, who serves as the executive director of the Arab American Association of New York (AAANY) and who was honored by former-President Obama's White House as a "champion of change," seems as if she makes it a point to attach herself to every social justice cause known to man and tie it to Palestine.
>
> Sarsour dismisses the fact that women in Saudi Arabia are treated as second-class citizens who are unable to drive, interact with men, and dress as they please as inconsequential.
>
> In addition to her dismissal of Saudi subjugation of women, she has attacked a documentary calling attention to the plight of women in the Islamic world. Sarsour has been a vocal critic of the executive producer of the film, Ayaan Hirsi Ali, a former Dutch Parliamentarian, ex-Muslim, and vocal critic of Islam who was also the victim of female genital mutilation.
>
> In 2011, Sarsour took to twitter and vulgarly berated Hirsi Ali; and ACT for America founder, Brigitte Gabriel, and said, "I wish I could take their vaginas away – they don't deserve to be women." This is especially vulgar considering the suffering Hirsi Ali has endured.[4]

# PREFACE

In 2015, German Chancellor Angela Merkel, the chief feminist-socialist leader in the West and the European Union, waving the feminine flag of compassion and altruism, declared an open door policy, inviting Muslim migrants from the distressed and war torn regions of the Middle East and North Africa. Her open borders policy, accepting an endless flood of migrants, came under severe criticism, as millions of Islamic people, mostly young men, were accepted into the nation, threating to change its demographic composition and culture. This led to a major social upheaval, with many sexual assaults and terror attacks taking place by Islamic migrants in Germany and other countries in Western Europe. Merkel defended this policy even in the wake of such attacks. She purportedly did not feel guilt about the attacks, stating that "even worse consequences" could have transpired had she not opened the country to these immigrants. The assailants, according to the Prime Minister, "wanted to undermine our sense of community, our openness and our willingness to help people in need."[5] In fact, in a somewhat bizarre attempt at responding to the sexual assaults against German women, the German Department of Health designed websites to *instruct* migrant men how to flirt with German women and even how to have sex with them—including detailed instructions with drawings! This, it would seem, is a pathetic example of state overreach into private lives, the socialist nanny state believing that instruction versus protection and law enforcement will magically solve a problem that is deeply rooted in cultural values and cultural clashes.[6]

The fact that Merkel has not been deterred in her immigration policy by this wave of terrorist attacks carried by Islamic immigrants on European soil—primarily in England, Germany, Belgium, and France—is viewed by her opponents as culture suicide. It seems that she is compelled by some force that belies her role as primary protector of her nation. And indeed she is. Her altruistic compulsion is driven by hormones that create the maternal imperative she seemingly cannot resist, even with her citizens being attacked in the most vile ways. In the summer of 2016, while she admitted that the continent was being "ravaged" and that attacks in Germany were "horrifying and depressing," she refused to let the nation shut out desperate refugees from the Middle East.[7] Merkel's character is the figure of the feminist ideal in contemporary Western culture—depicted in *Der Spiegel* in September 2015 as a nun, called "Mother Angella," personifying the feminine spirit of Christian love and altruism. She is a childless woman who pursues her career in politics for the "common good" of the people, rather than having a family of her own and children to love and bond with. Feminists, being against the traditional patriarchal family structure, who seek a career and independence, assuming a competitive, masculine posture, yet lack testosterone, the hormone of male competitiveness. Therefore, they are driven by their feminine, maternal instincts. The care and bonding hormone, oxytocin, drives them to seek compassion and bonding by unifying the entire world population in brotherly love and global harmony, a spirit best captured in John Lennon's song of the hippie generation during the sexual revolution of the 1970s: *Imagine there's no countries/It isn't hard to do/Nothing to kill or die for/And no religion, too/Imagine all the people/Living life in peace...*

And it's not only Merkel who has eschewed the natural inclination to bear children. In rather stunning contrast to monarchies and dynasties throughout world history that were compelled to carry on their lineage through their children, the 2017 leadership of almost all the major European nations are childless. Testosterone collapse leads to the death cycle of civilization as observed in these postmodern leaders not having progeny for the future (see Table 1).

| Germany | Angela Merkel | No children |
|---|---|---|
| United Kingdom | Theresa May | No children |
| France | Emmanuel Macron | No children |
| Italy | Paolo Gentiloni | No children |
| Sweden | Stefan Lofven | No children |
| Netherlands | Mark Rutte | No children |
| Scotland | Nicola Sturgeon | No children |
| European Commission | Jean-Claude Juncker | No children |

Table 1: 2017 European leaders without children

# PREFACE

We are witnessing Western culture moving to an extremely altruistic posture, losing its previous dominant status, and sacrificing its own existence in order to be taken over by foreign populations under multiculturalism. This cultural suicide, though it may seem insane and contrary to the biological imperative to sexually reproduce in order to propagate the genes, is actually deeply rooted in the origin and evolution of sex. In the following chapters, we shall explore this puzzle, discovering the nature of the evolved adaptations that gave rise to these biological forces that drive the rise and fall of human civilization.

# INTRODUCTION

In my first book, *The Objective Bible*, I explored the evolution of Western philosophy, analyzing how these ideas shaped our social and religious institutions throughout history. In my second book, *The Testosterone Hypothesis*, I built on my research in the history of philosophy, integrating neuroscience and the latest empirical research on our how brain structure shapes our personalities in order to further develop a new theory of how hormones frame our minds through the ages and drive our cultural evolution. In this work I wish to further expand on my previous research, through analysis of fundamental transformations in our culture, thereby expanding our interdisciplinary knowledge in the humanities, neuroscience, biology, and natural science. With these newly acquired tools, we can understand the forces that shape our cultural evolution and history, and we can work to gain independence as humans and be in charge of our own destiny.

## A Social Sex-Change Operation

Today in the West, we are in a period of profound change in social dynamics. Men are becoming effeminate, losing their cultural dominance, and the feminists, who steer toward socialism and submission to Islamic jihad of domination, are gaining power in our societies, taking prominent, leadership positions in academia, media, and the government. They are "asking" men to sit aside and quietly submit to their control and vision of the social good. This trend, of course, increases tension and social strife as the past order and structure of our social organization is experiencing rapid change and fundamental transformation. We are living in a time of increased social polarization and animosity between conflicting groups, our culture being divided and polarized across the lines of gender, race, nationality, and religion. There are many theories on the causes and drives of such great social upheavals. Some, like the philosopher and novelist Ayn Rand, believe history is driven solely by ideas, while others, such as the German philosopher Friedrich Hegel, believe in a *zeitgeist*, literally a spirit of the time that compels individuals to group together in collective waves of social mood, and subsequently reinforce the spirit of the time. Christian theology envisions a cosmic battle between good and evil, as portrayed between the spirit of Christ and the anti-Christ. Many more believe in vast conspiracy theories. On the left, the feminist and cultural Marxists believe there is a patriarchy that seeks to dominate and abuse them under the exploitive capitalist system, which they attribute to "toxic masculine dominance." While on the more masculine right wing, many believe in a global conspiracy for a "new world order" or a communist one-world government that seeks to subject humanity to its total control.

In reality all sides see elements of the truth; however, in this book, research into the biological roots of our group behavior reveals that the secret force driving the dynamism of human history is *testosterone*, the sex hormone that controls and shapes the sexual organization of society. It is a dominance hormone that drives men, among other mammals and social animals, to seek dominance and hierarchy structures and motivates them to reach for greatness.

As I explore in *The Testosterone Hypothesis*, similarly to how testosterone drives male erections, sex, and procreation, it can impel man to erect huge buildings, sky scrapers that reach for the heavens, or seek independence from a central religious authority as in the Protestant Reformation that broke from the Roman Papal authority of the Catholic Church. It can inspire men to declare a revolution against great kings, like the Founding Fathers of the United States did in 1776, or seek to control and dominate nature through the quest for enlightenment, as in the scientific and industrial revolutions.

However, hormone levels oscillate in up and down cycles. Just as they can carry humanity up in waves of rising social mood and euphoria, their fall can bring men back down into submission, as in the Dark Ages, when Western civilization collapsed from within, defenseless, to be conquered by the barbarians from without.

# INTRODUCTION

It seems that the period of the last four centuries, since the seventeenth century period of the Enlightenment, in which the West rose to world power and dominance, is coming to an end. The Western male was the global alpha male, the dominant force in the rise and spread of Western science, industry, economic, and political influence around the world. Among the great European powers, the leading force in the rise of West was the British Empire, which ascended to prominence, establishing a world empire from India and Australia in the Far East to colonizing Africa in the south, and taking over North America in the West. This spread of Judeo-Christian civilization and the capitalist system gave rise to vast wealth, critical scientific discovery, and prosperous industry on an unparalleled level never seen before in human history.

Nevertheless, as much as these achievements have caused a sense of invincibility at our peak point of exuberance, the forces that drive human history and cultural evolution never cease to work. They can plummet us down as dramatically as they have lifted us up. It seems that the West has been gradually losing its drive for expansion and its motivation for greatness and is now contracting and shrinking from its former power position and days of glory. It can even be said that it's forfeiting its former self-confidence and willingly accepting its own defeat. Every tenet of the former rise of the West, from its masculine leadership to its Judeo-Christian religious traditions, to its political and economic system of capitalism and individual rights and the rise of science and industry, is now attacked. This attack comes primarily from the feminist–socialist controlled academic institutions, evil relics of the past, ironically associated with "toxic male" dominance of the culture.

The fall of the West from dominance and an expansive position into submission and diminishment suggests that these cycles of civilizations' rise and fall might prove to be driven by dominance hierarchies among individuals in human and mammal societies. These hierarchies are reinforced even at a physiological level. Social psychologist Amy Cuddy argues that merely posing in the postures powerful people tend to choose, those that are "open and expansive," causes changes in endocrine levels and in behavior:

> High-power posers experienced elevations in testosterone, decreases in cortisol, and increased feelings of power and tolerance for risk; low-power posers exhibited the opposite pattern. In short, posing in displays of power caused advantaged and adaptive psychological, physiological, and behavioral changes, and these findings suggest that embodiment extends beyond mere thinking and feeling, to physiology and subsequent behavioral choices. That a person can, by assuming two simple 1-min poses, embody power and instantly become more powerful has real-world, actionable implications.[1]

With the decline of testosterone, which is also associated with other happy chemicals in the brain, such as serotonin, the status hormone, and dopamine, the motivation hormone, Western male dominance and the status of Western civilization itself is now in rapid decline, balancing on the verge of total collapse.

One need not look any further than the Top 40 chart to see evidence of an era's prevailing social mood. The 1984 anthem "I Need a Hero," by Bonnie Tyler, depicts a woman seeking a strong, heroic, traditionally masculine man, representing the wave of rising testosterone throughout the 80s and early 90s:

> *Where have all the good men gone/and where are all the gods?*
> *Where's the streetwise Hercules to fight the rising odds?*
> *Isn't there a white knight upon a fiery steed?*
> *Late at night I toss/and I turn/and I dream of what I need.*
> *I need a hero. I'm holding out for a hero 'til the end of the night.*
> *He's gotta be strong/and he's gotta be fast/and he's gotta be fresh from the fight.*

# INTRODUCTION

By contrast, the more depressed environment of the new millennium, especially post-2008, is represented in the confused lyrics of the popular song "The Fear" of 2009, in which seeking prosperity is mocked and meaning in life is elusive:

*I want to be rich and I want lots of money*
*I don't care about clever I don't care about funny*
*I want loads of clothes and f---loads of diamonds*
*I heard people die while they are trying to find them*
*I'll take my clothes off and it will be shameless*
*'Cause everyone knows that's how you get famous*
*I'll look at the sun and I'll look in the mirror*
*I'm on the right track yeah I'm on to a winner*
*I don't know what's right and what's real anymore*
*I don't know how I'm meant to feel anymore*
*When do you think it will all become clear*
*And I'll be taken over by the fear*

Following the financial collapse and severe global recession of 2008, Western culture has experienced extreme depression in social mood. This depression is evidenced quite literally by the increase in antidepressant drug use of people aged twelve years and older. The rate of antidepressant use grew 400% in the decade preceding the financial collapse.[2] Moreover, changes in the culture's sexual organization have been obvious in years just preceding the financial crash, and certainly thereafter. I term this as a social "sex change operation." This decline in social mood can be seen readily in how the stance on same-sex marriage "evolved," focusing particularly on national leaders Barak Obama and Hillary Clinton. Whereas Obama, in 2008, stated, "I believe that marriage is the union between a man and a woman...for me as a Christian, it's also a sacred union; God's in the mix," by 2010 he admitted that his views were changing. And by 2013 he called the US Supreme Court ruling against individual states' right to ban same-sex marriage a "victory for America." Likewise Hillary Clinton, when campaigning for the US Senate in 2002, gave a definitive "no" when asked if New York State should recognize same-sex marriages. A decade later, she resoundingly supported the Court's decision, calling it an "historic victory for marriage equality and the courage and determination of LGBT Americans who made it possible."[3]

This sex-change operation of the culture has manifested in complicated ways. Just prior to the publication of this book, in the fall of 2017, the US popular culture was rocked by the sexual harassment and assault scandal. Beginning with Hollywood mogul and Democrat Party mega fund-raiser Harvey Weinstein, the sexual abuse and manipulation of women that had transpired for decades as an "open secret" was revealed in *The New York Times* and *New Yorker* magazine articles. After the initial accusations against Weinstein were published, the floodgates opened, and man after man in entertainment, news, politics, and academia was brought down with descriptions of everything from lewd comments to outright rape—of both women and young men. What makes this complicated is the kind of schizophrenic nature of the dynamic. It seems that an outcome of the sexual revolution—free sex and flaunted sexuality—is at odds with our true biological character as women and men. The whistle blowing on the one hand indicates a good thing, in that men should not be abusing and harassing women, and for some reason, the time was ripe for this to be taken seriously. However, the feminists don't seem to see that they are requiring men to act chivalrously while women are supposed to be able to act out in any sexually provocative ways they want to.

In fact, the dominating feminist culture that vilifies male sexuality has made a play to gain exactly this kind of control. Conferring on all males an intrinsically evil and exploitative nature, feminists have ushered in a new Dark Age doctrine of sex as Original Sin. A *New York Times* opinion piece entitled "The Unexamined Brutality of the Male Libido" portrays the issue like this:

> For most of history, we've taken for granted the implicit
> brutality of male sexuality. In 1976, the radical feminist
> and pornography opponent Andrea Dworkin said that

the only sex between a man and a woman that could be undertaken without violence was sex with a flaccid penis: "I think that men will have to give up their precious erections," she wrote. In the third century A.D., it is widely believed, the great Catholic theologian Origen, working on roughly the same principle, castrated himself.[4]

Professor Laura Kipnis laments that "once again" nineteenth-century ideals based in "gendered nature" require women to shoulder the responsibility of taming men's out-of-control sexuality.[5] Echoing many of her sisters in the culture, she is clearly not happy with that prospect:

And we're officially back in the 19th century, where exactly such ideas about "gendered nature" held sway. Men are naturally predatory, thus it's women's task to reform them with our instinctive femininity. Aargh. (From 1856: "It is woman's womanhood, her instinctive femininity, her highest morality that society now needs to counteract the excess of masculinity that is everywhere to be found...")[6]

But the women who only months before marched on Washington to protest President Trump aren't having it. Characteristic of the sexual harassment scandal is that the liberated woman is not going to give up any of her hard-fought sexual or professional freedoms. Their insistence is that the men alone change.

We are now in the latter phase of cultural collapse, in which the feminists are even turning against the men in their own left-wing establishment: Harvey Weinstein, Bill Clinton, Charlie Rose, Kevin Spacey, etcetera. Such a phase is similar to Stalinist purges, in which people faced trial without due process and were sent to the labor camps or gulags. In the current cultural climate of the US, the allegations are adjudicated by the court of public opinion, and the gulag the men are condemned to is their career's end. This is not to say that there should not be consequences for actual abuse! But with no set standard to establish what constitutes abuse and no clear path for due process, the atmosphere has taken on the quality of a witch hunt, in which only an outcry is needed to ruin a man's reputation, with no burden of proof.

## Feminists and Islamists: Strange Bedfellows

The post-traumatic social environment after the depressive stage of 2008 has also led to the emergence of a very troubling phenomenon coming from the Middle East: the rise of an extremely violent, sadomasochistic strain of Islamic jihad that glories in rape, murder, beheadings, and mass executions and enslavement. Variously referred to as ISIS or ISIL, the Sunni group claims itself a worldwide caliphate with authority over all Muslims and self-identifies as the Islamic State (IS):

> There are many reports of sexual abuse and enslavement in ISIL-controlled areas of women and girls, predominantly from the minority Christian and Yazidi communities. Fighters are told that they are free to have sex with or rape non-Muslim captive women. Haleh Esfandiari from the Woodrow Wilson International Center for Scholars has highlighted the abuse of local women by ISIL militants after they have captured an area. "They usually take the older women to a makeshift slave market and try to sell them. The younger girls ... are raped or married off to fighters," she said, adding, "It's based on temporary marriages, and once these fighters have had sex with these young girls, they just pass them on to other fighters."
>
> The capture of Iraqi cities by the group in June 2014 was accompanied by an upsurge in crimes against women, including kidnap and rape. According to Martin Williams in *The Citizen*, some hard-line Salafists apparently regard extramarital sex with multiple partners as a legitimate form of holy war, and it is "difficult to reconcile this with a religion where some adherents insist

that women must be covered from head to toe, with only
a narrow slit for the eyes."[7]

As ISIS gained territory through a catastrophic wave of violence
and terror, another troubling cultural phenomenon has taken place,
primarily in the West. This is the popular trend among feminists
finding their sexual fantasies and arousal through the novel and movie
*Fifty Shades of Grey* and the sequel, *Fifty Shades Darker*. These
"romance" stories are notable for their explicitly erotic scenes that
feature elements of such sexual practices as dominance/submission,
sadism/masochism (BDSM), and bondage/discipline. The book was
first published in 2011, becoming the top best-seller all around the
world and selling more than 130 million copies by 2015.

Why are feminists allured by such sadomasochism?

It is testosterone collapse that leads to the spread of the feminist-
socialist sadomasochistic culture because of sexual depression,
repression, and aggression. This was amplified more than two-
hundred years ago in the writings of the Marquis de Sade:

> De Sade held no illusions about the natural goodness of
> man, but he believed that with complete economic and
> sexual equality human conditions could be greatly
> bettered. He went far beyond the 'advanced' social
> thinkers of his time and even of the present day. One of
> his earlier biographers Geoffrey Gorer, who wrote *The
> Revolutionary Ideas of the Marquis De Sade*, pointed out
> that Sade was in complete opposition to contemporary
> philosophers for both his "complete and continual denial
> of the right to property," and for viewing the struggle in
> late 18th-century French society as being not between "the
> Crown, the bourgeoisie, the aristocracy or the clergy, or
> sectional interests of any of these against one another,"
> but rather all of these "more or less united against the
> proletariat." Gorer argued, "he can with some justice be
> called the first reasoned socialist."[8]

*The New Dictionary of Cultural Literacy* defines the expression
"It takes two to tango" as certain activities that cannot be performed

alone–such as quarrelling, making love, and dancing the tango. These actions are the product of mutual cooperation, which carry either a positive or negative connotation. As we shall discover in this research, the simultaneous rise of ISIS, Islamic jihad coming from the Middle East, and the feminism, sadomasochism, and a culture of submission in the West is no coincidence. This is rather a well-orchestrated sequence of events, all globally set in motion and guided by the demon-like powers of our complex biological drivers when stress hormones become imbalanced to an extreme—our sex hormones, synchronizing world events by transforming the sexual organization of our societies.

In 2010 a revolutionary wave of violent demonstrations, starting in Tunisia and spreading across the Arab world began. Civil uprisings and power struggles resulted in large scale conflicts in many Muslim countries such as Libya and Syria and led to the flood of refugees streaming into Europe. Many socialists on the left were initially very optimistic about a wave of democracy sweeping the Islamic world, and donned the movement "The Arab Spring;" but the result of the upheavals was actually the ascendency of extremist Islamic jihadi militants, such as ISIS and the Muslim Brotherhood.

## Worldwide Evidence of Sex Wars

And it's not only the Middle East or America that give evidence of the cultural devastations of testosterone collapse. To demonstrate the global existence of this phenomenon, we shall explore its many manifestations all around the world. *Men Going Their Own Way (MGTOW)* is a counter-feminist movement that is growing in recent years in the West as the sex wars intensify. This movement calls for more men to abandon traditional gender role expectations such as marriage and having children, and to not be abused by the feminist controlled legal and judiciary system that disenfranchises men.

In Japan, in 2015, the following headline appeared regarding Japan's diminishing sex drive and collapse in the family structure: *Japan's Population Time Bomb Worsens as 'Herbivore' Men not*

# INTRODUCTION

*Interested in Sex Extends to Married Males.* The reason behind such pessimism is that fewer Japanese are having sex. The phenomenon of men lacking interest in sexual engagement has led to the term *soshoku-kei danshi*, or herbivore men, to describe them. However, what is quite alarming is that it is felt not only among single males, but also among married men.[9] China, in 1979, introduced its one-child policy for the purpose of population control. While this policy has gradually been relaxed since 2013, because the communist authorities became concerned about the aging population and the implications for economic growth, testosterone collapse has caused a weak response to the new governmental policy:

> But many are not interested, and the response has been lackluster, despite an initial burst of built up demand. China now has the lowest fertility rate in the world—1.05 according to China's 2016 State Statistical Bureau data and reported by Liang Jianzhang and Huang Wenzheng in a recent *Caixin* article.
>
> China's skewed sex ratio at birth, officially at 120 boys for every 100 girls born (but as high as 143 to 100 in rural areas for second births), exacerbated a long tradition of son preference and has created a generation of extra men.[10]

Aligning with this low fertility rate is the concern among Chinese authorities about the "masculinity crisis" in their nation. It is recognized that the increase in "effeminate men" might impact negatively the nation's security, "because it reflects powerlessness, inferiority, feminized passivity, and social deterioration reminiscent of the colonial past when China was defeated by the colonizing West."[11]

In Europe, Germany and Sweden, as well as other nations, show indications of diminished birth rates resultant of lowered sexual appetites and indicative of declining testosterone. Not only has the birth rate in Germany fallen by eleven percent since the year 2000, according to Europe's Population Policy Acceptance Study, some twenty-three percent of German men answered that the perfect number of children was "zero."[12] Media reports in Sweden, notorious for its libertine sexual ideals, claim that interest in sex has lessened, based on surveys such as a 3,000 person questionnaire conducted in 2013. The government is concerned enough about this trend to have launched a multi-year study that will culminate in 2019. Its minister of public health recognizing that "if stress and other health issues are affecting Swedes' sex lives, 'that is also a political problem.'"[13]

At the same time that these nations identify declining sexual and reproductive drive, Sweden, considered the most feminist and most liberal European nation, like Germany, is accepting unprecedented numbers of Islamic immigrants. And, like Germany, has suffered consequences that has made it the *rape capital of the West,* even leading worldwide:

> Forty years after the Swedish parliament unanimously decided to change the formerly homogenous Sweden into a multicultural country, violent crime has increased by 300% and rapes by 1,472%. Sweden is now number two on the list of rape countries, surpassed only by Lesotho in Southern Africa.
>
> In an astounding number of cases, the Swedish courts have demonstrated sympathy for the rapists, and have acquitted suspects who have claimed that the girl wanted to have sex with six, seven or eight men.[14]

It should be noted that the vast majority of the incoming stream of immigrants is not composed of families, but of single men, unaccompanied by extended family. *Politico* ran a headline that read, *Europe's Man Problem: Migrants to Europe Skew Heavily Male— and That's Dangerous.* The article goes on to say, "In fact, the sex ratios among migrants are so one-sided—we're talking worse than those in China, in some cases—that they could radically change the gender balance in European countries in certain age cohorts."[15]

This stream of apparently disparate events, from the decline in sexual interest and subsequent birth rates, to the growth of militant Islam in the Middle East, to the rise of militant feminism in the West, can be interconnected as we look at our social structure through the prism of our biological impetus, our hormones. We have to connect the dots to discover the fundamental concepts that unify these trends and these doctrines. Both, Islam, which means submission, and the feminist–socialist and environmentalist credo that seems to be taking over the Western mind, seek to bring men into submission to an authoritarian ideology and end the productive era of individual rights, freedom of speech, and capitalism that the West embraced since its rise from the Enlightenment. The modern world has been fundamentally called into question and attacked, as the postmodern era subjugates with deconstruction and demolishing ideologies.

In the early part of the twentieth century, the rise of authoritarianism flourished with communism in the Soviet Union and Maoist China, Nazi-Socialism in Germany, and Fascism in Italy and Japan. Finally such statism was curtailed—in the West at least—after WWII and in the Cold War, with the values of classical Western liberalism on the winning end. This euphoric triumph for the West led Francis Fukuyama to publish *The End of History and the Last Man* in 1992, arguing that the superiority of Western liberal democracy might mark the end times in humanity's sociocultural evolution, reaching its synthesis, or final point, in the Hegelian sense. Mitigating the overwhelming optimism in Fukuyama's prophesies, Samuel P. Huntington published *The Clash of Civilizations and the Remaking of World Order* in 1996, stating that cultural and religious identities will continue to be a major source of conflict, particularly on the border lines between cultures, singling out Islam, described as having "bloody borders." Huntington explained that that the difference between civilizations, particularly Islam and Christianity, is too great, as a product of many centuries' historical and cultural development, and that these disparities would bring more waves of tension in the future.

The ebullient social mood following WWII peaked in 2000. For decades Western prosperity spread across the globe; yet troubling undercurrents have been gradually brewing, largely unnoticed, beneath the surface. Islam has been radicalized since the 1970s, and even in the West, family values have been declining. The great recession after 2000 and the tragic attack on the World Trade Center in September 2001 by Islamic jihadists on a suicide mission, has signaled that a long period of positive social mood has come to an end, and the tides were now shifting in the opposite direction.

The profound change in the global attitude toward themes associated with sexuality and religion is evident in the transformation experienced in the Muslim world since the 1950s, when secular regimes were in power. However, this region has gradually endured greater radicalization by radical Islamist ideology, reminiscent of the Dark Ages, with sexual repression, depression, and aggression toward women, finally leading to their submission through the traditional ultra-conservative covering of the entire body and head with the burka. In a video from Egypt of a speech President Nasser gave in 1958, Muslims laughed at the Muslim Brotherhood's idea to force the hijab on women, calling on the religious clerics to wear the covering themselves.[16]

The negative trend in social mood that resulted in an abusive culture was not confined to Islamic culture. In 1979, *Caligula*, an Italian-American pornographic film that was wildly controversial because of its explicit sexual content, but nonetheless quite popular, was produced, depicting the rise and fall of the Roman Emperor Caligula. Caligula was an insane, morally depraved emperor, ruling Rome in the period 37-41 CE. He was infamous for indulging in wild orgies and sexual perversions, killing people on a whim, causing starvation, and commanding to be worshiped as god at the temple in Jerusalem. Similarly the Italian film *Salo*, or *120 Days of Sodom*, featured in 1975, is commonly listed among the most controversial films ever made. It is based on the sadomasochistic, pornographic novel, *120 Days of Sodom*, written in 1785 by the Marquis de Sade before the period of the bloody French Revolution. The film presents the story of four wealthy libertines who kidnap a harem of a few dozen teenagers and engage in orgies of sexual abuse and torture, intensifying in cruelty until culminating in the slaughter of the victims.

# INTRODUCTION

## Low Testosterone Affects Perspectives

A further testimony to the power of testosterone collapse shaping ethical principles about human sexual desires and attitudes about our faculty of reason is a particularly postmodern reading of the Kantian understanding of sexuality from the 1780s. In 2016, Raja Halwani, a professor of philosophy at the School of the Art Institute of Chicago, and author of *Philosophy of Love, Sex and Marriage* (2010), published an article condemning sexual desire as merely "objectifying" under his interpretation that Kantian ethics deny the body:

> The 18th-century philosopher Immanuel Kant believed that human beings tend to be evil. He wasn't talking about some guy rubbing his hands and crowing with glee at the prospect of torturing an enemy. He was thinking about the basic human tendency to succumb to what we want to do instead of what we ought to do, to heed the siren-song of our desires instead of the call of duty. For Kant, morality is the force that closes this gap, and holds us back from our darker, desiring selves.
>
> Once desire becomes suspect, sex is never far behind. Kant implicitly acknowledged the unusual power of sexual urges and their capacity to divert us from doing what is right. He claimed that sex was particularly morally condemnable, because lust focuses on the body, not the agency, of those we sexually desire, and so reduces them to mere things. It makes us see the objects of our longing as just that – objects. In so doing, we see them as mere tools for our own satisfaction.[17]

To the degree that modern thinkers understand mankind as unable to govern sexual impulses with reason, they strip individuals of their God-given agency to act in their own best interest. This led in the early twentieth century to the rise of political ideologies and social constructs that denied individuality and defined people groups as biological "superorganisms." In a period of testosterone decline, we are inclined to lose our individualism and independence and group together into tighter social units that fight each other for a higher position in the social dominance hierarchy. *Biopolitics* was a term coined in 1911 pertaining to the intersectional discipline of biology and politics, developed by Rudolf Kjellén (1864–1922). It is an organismic view of society and state as a "super-individual creature." Kjellén asserted that biological factors underlie "the civil war between social groups" that comprise the state.[18] The Nazis accepted such an organismic view of society, rooted in biology, and used the concept in reference to the biological origins of their nation in race and genetics.[19]

As I shall address in the following chapters, this brand of synthesizing biology and social sciences is extremely problematic and dangerous. It was embraced by racist, fascist ideologies and regimes in a period of extreme negative social mood, the 1930s, that viewed man as an animal lacking free will and reason and motivated by animalistic social instincts for group wars. With the troubling history of Darwinian evolution adopted by the racist Nazi regime in the twentieth century in its eugenics program for racial purity, we can see how perilous it is, both on the right and left of our political spectrum, to prefer an idealistic position based on such transcendent ideals for the "social good" that chooses to ignore biological factors that determine our social structure.

One important factor to consider is that both fascist and communist regimes promote group conflict by emphasizing the *differences between us*, usually based on race, social class, or religion. However, nature creates diversity in order to promote specialization. The emergence of complexity through the division of labor in biological and human societies is similar to how the cells in our body differentiate into many group functions, such as heart cells, blood cells, and brain cells. Hence, our genetic and social diversity should be celebrated as the enabling factor that promotes the incredible complexity of the advanced stage of civilization we have reached, with global trade and economic and scientific exchange for mutual benefit, uplifting and enriching humanity across ethnic and cultural borderlines.

# INTRODUCTION

The significant message of this book is that our culture wars are primarily driven not by our different genetic makeup, but by periodic hormonal cycles that are shared by the entire human species, which oscillate between periods of positive social mood, rising individualism and peaceful cooperation, and depression and group cohesion that lead to social conflict and war. Rather than focusing on the current forces that set us apart, we should consider ourselves as part of a long lineage of one human family that is together facing the cyclic forces that nature has bestowed upon us, going through the ages in the course of our common history and our shared humanity. By focusing on man versus nature, rather than men against each other, we can join forces in our common struggle to improve our state of life. If we put our mind to it, as the rational man, we can dominate nature by discovering its workings, rather than being divided and conquered by it. As Ayn Rand declared in The Fountainhead, "The creator's concern is the conquest of nature. The parasite's concern is the conquest of men."

## A New Model for the Social Sciences

As we shall expand on later, the evolving cycles of history are regulated by our ancient mammalian brain in order to adapt our social superorganism to changing environmental conditions. These are primarily the changing energy levels incoming from solar radiation, affecting global climate change, and thus, plant growth and the entire animal food chain. However, humanity has reached a technological level of sophistication and scientific knowledge today that can enable us to steer our course with modern tools of reason, rather than be guided by primeval impulses rooted in our more primitive evolutionary past. We shall further study the evolutionary drivers for group conflict throughout this work to clarify their origins and promote a rational discourse regarding the substantial challenges we face.

Humanity will have to contend with these difficult philosophical and neuroscientific problems of mind versus body, rather than ignore them, as we are inclined to do today. These issues will continue to haunt us, due to the evolutionary, animalistic origin of our brains and how it shapes our human mind—whether we like it or not. In the process of fighting each other in the propaganda wars, our society is becoming more divided and fractured into competing social groups in a Darwinian struggle for social dominance and submission. We are losing sight of objective, scientific discourse and rational debate. This will only make us weaker and allow for the forces of degeneration and entropy to consume our powers. Our hope for improving our lives is through returning to the rational and moral principles that have guided Western civilization since the Enlightenment.

Hence, I will suggest here that we choose to contend boldly with our evolved psychological adaptations, study their workings, and use this knowledge to command our own human nature by learning to master our animalistic impulses. This is the only rational path that humanity can take if it is to commit to creating a moral and civilized culture as prescribed by our common Judeo-Christian heritage that promotes the values of human life and liberty.

The purpose of the information I present here is to establish the evidence for a new scientific understanding of the biological origins of our great social transformations. Without any unifying scientific model to understand the myriad events that drive our social world, we cannot form a coherent understanding of the forces that drive our history, culture, and social life. Lacking the kind of systematizing approach in the social sciences that we have formed in the natural sciences, such as physics and biology, we are left dumbfounded and confused by the endless stream of events that seems to be flowing in a completely random and chaotic flux. By discovering the orderly nature, in periodic cycles, of the recurring themes in our history, we can revolutionize our comprehension of the social sciences like Newton, with his theory of classical mechanics, produced in the physical sciences, and Darwin, with his theory of evolution, brought to the biological sciences. This will complete the great project of the scientific revolution, which the West has taken on since the Age of Enlightenment: to live by a vision of the rational man in command of

# INTRODUCTION

nature and his own mind. Harvard biologist E. O. Wilson, in his 1998 book, *Consilience: The Unity of Knowledge,* has called for a similar vision of synthesis between the humanities and natural sciences.

In the late seventeenth century, Isaac Newton, the father of modern science, led the scientific revolution by his groundbreaking assertion that our physical universe, rather than being manipulated by multiple, inexplicable forces and spirits, is a mechanical operation, a clockwork model, being held together by the single, principle force of gravity. Furthermore, gravity, the same force that governs the course of celestial bodies in rotational orbit around the sun, causes the fall of an apple here on earth. The rise of Newtonian mechanics, with its clear, orderly model, has enabled humanity to grasp our world according to a scientific method and to command nature, ending the age of irrational, magical delusions that rendered man helpless and in submission to forces he cannot comprehend.

Similarly, in the nineteenth century, Charles Darwin posited the theory of evolution. His theory stated the single principle of natural selection for the origin of all species, creating a clear, scientific basis for understanding how a variation of species arises in the world of biology. Until Darwin's publication of *On the Origin of Species* (1859) it was believed by the academic establishment that all species were created differently, with no one principled process taking place. Darwin shattered the mystical view of creation with a new understanding of creation as an evolutionary process. In relation to the humanities and the social sciences, the pre-Darwinian, anachronistic point of view still pervades the academic world. My series of books is designed to shatter this simplistic view of our cultural origins with a new, groundbreaking thesis that the origin of cultures is also an evolutionary process, similar to that of the biological world. Understanding the complexity of human civilization as a dynamic process will revolutionize the humanities and propel us to the next level of scientific knowledge.

In *The Testosterone Hypothesis* I laid the foundation for this revolution in the social sciences by describing the primary role of hormone cycles in our cultural evolution. However, there is still much work to be done to further explain the biology of sex. The rise of dominance hierarchies in the evolution of sexual reproductive strategies is a method of creating complexity and genetic diversity in the biological world. This happens through phase transitions between periods of established order and hierarchy periodically punctuated by catastrophic chaos and turmoil, in which new synergies are created and different life forms take shape.

Now, in the twenty-first century, we have the opportunity to bring the unifying methods of Newtonian mechanics and Darwinian evolution into the social sciences with a new vision that dares to challenge long-standing, mystical traditions that threaten our modern civilization and way of life.

# 1

# ON HORMONES, FEELINGS, AND IDEAS: MIND–BODY DUALISM

*A new Sex is the point of contact between man and nature, where morality and good intentions fall to primitive urges. I call it an Intersection.*[1]
*Judaism, Christianity's parent sect, is the most powerful of protests against nature. The Old Testament asserts that a father god made nature and that the differentiation into objects and gender was after the fact of his maleness.*[2]
— Camille Paglia

The underlying forces that drive our social life, whether for greater harmony and peace or to social strife and war, are sex hormones that regulate the sexual organization of society. To understand how biology, genetics, and the bioenergetics of life play into the complex systems of human cultures ultimately will equip us to transcend our biological nature—the nature that leads inevitably to death. These systems are designed by evolution to change cyclically: complex societies gain independence, develop social order with political and religious institutions, only to gradually disintegrate from within, collapse, be conquered by invasions from the outside and necessarily assimilate. We must study these natural rhythms of history and recognize what is happening in order to create a better future for the human race.

Hormonal levels, such as the sex hormones and dopamine, the motivation hormone, drive men to form social structures according their sexual motivation. This shapes the entire worldview of the community as well as the social environment, impelling the group to code its ideology into religious structures and organizations that formulate the underlying biological impulses into symbolic archetypes of meaning. In this chapter, we will look at some ideas about how and why humans have organized themselves relative to these hormonal waves and social moods throughout history and what we can recognize as outcomes of our society's current state of mind.

A significant premise for this work is that levels of testosterone have fallen in the past twenty years.[i] A Finnish study conducted in 2005 affirmed other international studies in showing that testosterone levels dropped twenty percent from a generations earlier.[3] Moreover, while high testosterone is often mischaracterized as merely the "rage and rape" hormone, its reality is much more nuanced. James Dabbs, the author of *Heroes, Rogues, and Lovers: Testosterone and Behavior*, writes that since before recorded history, testosterone's importance has manifested in all forms of relationships, affecting both men and women in their cognition, such as with spatial relations, as well as their linguistic abilities.[4]

High testosterone makes men feel more confident, optimistic, and prepared to create a family.[ii] It spurs men on to take parental

---

[i] For a more detailed discussion of the human endocrine system and the role of testosterone in particular, please refer to my previous book, *The Testosterone Hypothesis* (2015).

[ii] There is interesting research indicating that men with higher testosterone level are more likely to marry than men with lower testosterone. Testosterone drives men to create a family, but later it declines as they care for their children with the feminine quality of empathy. Thereby testosterone regulates the entire reproductive life cycle. Peter B. Gray, "The Descent of a Man's Testosterone," *Proceedings of the National Academy of Sciences of the United States of America* 108, no. 39 (September 2011): 16141–16142, doi:10.1073/pnas.1113323108.

responsibility as fathers and to form a society based on masculine virtues of potency, power, rationality, and creativity. Such confidence equips men to take control and exercise dominance over the natural world. In periods of increasing testosterone the masculine ideal and image carries the culture to greatness—that is, to stand tall in a dominant, expansive posture, commanding nature and the material world and looking up toward the skies, to the Father in heaven, the Creator figure of Genesis. On the contrary, however, when society experiences a wave of decline in this sex hormone, social mood turns negative, it shrinks down toward our material origins and surrenders to fate, submitting to the destructive forces of Mother Nature. Men feel impotent, weak, and submissive, oppressed under a fatalistic worldview that condemns them to their fate.[iii]

Fundamental to these different states of mind is how they shape our very perceptions of self and the world we live in. From personal efficacy to societal prosperity, the tenure and tone of information promulgated depends on the social mood hormonal levels are creating. Low testosterone renders a perplexed state in which man feels utterly ruled by subjective emotion, the "animal passions" within. In this work, I seek to increase our recognition and understanding of the gap between the beast within us and our moral character as humans. In the modern age this was coined "the mind–body gap," the gap between our animal roots as the evolved driver of our psychological and social impulses and our moral and religious sentiments observed in the complex philosophical systems only humans can create. Yet, long before Renee Descartes investigated this gap, the prophet Isaiah exhorted the Israelites to "Come, let us reason

---

[iii] A study by psychologists at Swansea University's College of Human and Health Scientists has found that people's personal 'locus of control'–the extent to which we believe we can control events that affect our lives—is partly influenced by the amount of the 'male' hormone testosterone that we are exposed to in the womb. Gareth Richards, Steve Stewart-Williams, Phil Reed, "Associations between Digit Ratio (2D:4D) and Locus of Control," *Personality and Individual Differences* 83 (September 2015): 102-105, https://doi.org/10.1016/j.paid.2015.03.047.

together," echoing the primary faculty given man by God, that of reason over base impulses.

## Introduction to Dualism

We are all conceived as females. It is the increase of the male sex hormone, testosterone, starting during the sixth week of pregnancy, that creates the male body and brain characteristics in the fetus. One can chart the development of a male from his feminine origins to his full masculinity and through it explain the entire history of the rise of fall of Western civilization. In the past hundred-and-fifty years, this can be seen in the raging debates between the Judeo-Christian worldview with a divine Creator of an orderly, moral, and benevolent universe and the Darwinian, evolutionary view of a brutal and savage process of change that requires mass extinction in the barbaric war for the survival of the fittest. Humanity is in the grip of the forces of nature, but when at its greatest, it strives to overcome nature and command it.

According to our Western philosophical heritage, as humans, we have a dual existence as both biological animals that are driven by instincts for survival and propagation of our species and spiritual beings that possess the faculty of reason, wherein the mind is deemed as a living in a transcendent dimension, beyond our physical body. This dual philosophy of existence, conceived of by Plato, still holds firm in our consciousness—even in the age of science and technology that emphasizes empirical evidence and attacks ancient religious traditions. The struggle between the mind and body has always been the defining theme in human philosophy and religion. How do we reconcile our consciousness and the life of the soul with the material body destined for degeneration and death, trapped in the material world? In the history of Western civilization, our millennia old Judeo-Christian heritage encompasses the entire spectrum of different interpretations of this complex mind-body relationship.

The father of Enlightenment and modern philosophy, Rene Descartes (1596–1650), wrote his famous phrase, "I think, therefore I am." Asserting thought as the locus of identity, Descartes is credited with solidifying the mind–body distinction in the annals of Western philosophy. He argues that the nature of the mind (i.e., thinking) is utterly different from that of the body (i.e., the non-thinking thing); hence the one can actually exist without the other. Since the Enlightenment, the Cartesian worldview has been instrumental in developing Western science and philosophy in a view called the mechanical universe. Herein nature is deemed a clockwork machine made of particles in motion, moving in deterministic patterns, thereby discernible to the human mind. Moreover, in this Cartesian view, man's mind has retained its free will by separating itself from the confinements of the material world and elevating it to the realm of the divine, or the spiritual. Thus, over the last four centuries, Western philosophy has been residing in the best of both worlds: the mechanical universe is controlled by predictable natural laws, with forces we can study with our rational minds, yet we are able to perceive our mind itself as immaterial, not limited by the restrictions set by the mechanical principles that govern the rest of the mechanical universe.

In his scientific worldview Descartes asserted that the animal world, including humans, are utterly mechanistic. He understood that biology was also mechanistic in nature:

> I should like you to consider that these functions (including passion, memory, and imagination) follow from the mere arrangement of the machine's organs every bit as naturally as the movements of a clock or automaton follow from the arrangement of its counter-weights and wheels.[5]

However, his view of a mind–body dualism has been challenged in the last few decades by postmodern philosophy as well as a new scientific understandings in neuroscience. Neurologist António Damásio offers interesting insight to the question of the mind–body dualism in his 1994 book, *Descartes' Error: Emotion, Reason, and the Human Brain.*[6] Damásio's "somatic marker hypothesis" posits an emotional mechanism that directs behavior and decision making. If rationality is motivated by emotional signals, Descartes' dualist separation of rationality and emotion does not stand. This is Decartes' "error." Damásio asserts that "gut feelings" play a major role in navigating our crucial decision making—intuitive signals from limbic-driven surges emanate from the viscera. These Damásio calls "somatic markers." Rather than overriding rationality, somatic markers "direct attention towards more advantageous options, simplifying the decision process."[7] The emotive process serves as a filter in the midst of complex stimuli to aid the neocortex in assessing situations.

> Although [complex systems] differ widely in their physical attributes, they resemble one another in the way they handle information. That common feature is perhaps the best starting point for exploring how they operate.[8]

A theory like Damásio's might offer insights to how our brains function, but just as important is his focus on *complexity*. Biochemical interactions that make up the human mind, including sensory assessment, memory, and "intuition" combine as a highly complex system. Where Hume's understanding, or even Plato's long before him, was limited by a lack of perspective, such as evolutionary biology, subsequent advances in understanding about biology, chemistry, neuroscience, complexity and chaos theories, and evolution have opened the door to integrative theories like the somatic marker hypothesis of Damásio. Furthermore, such comprehension of the individual brain leads us to better comprehend our evolved trait of consolidating into "group brains," cohesive social groups that enable us to exponentially enhance our productivity and creativity in human culture.

## Organizing into Groups

There are many biological systems studied today in the emerging science of complex systems. Ants cooperate through division of labor to form a nest. Neuronal networks in our brain are composed of billions of neurons that form thousands of interconnections between them; and in a similar fashion we humans form social networks to advance our goals through group dynamics. *Swarm intelligence* is an expression given for the collective behavior of decentralized, self-organizing, complex systems. This concept is employed in artificial intelligence, but its inspiration comes from biological systems, such as animal herding, ant colonies, bee hives, bird flocking, bacterial growth, microbial intelligence, and fish schooling. These systems are composed of a population of multiple agents following simple rules of interaction between them and with their environment. The cumulative action of many agents, following simple rules, results in the emergent phenomenon of intelligent, complex global behavior not known to the individual components.

The great Enlightenment philosopher Francis Bacon's phrase, "Nature, to be commanded, must be obeyed," beautifully captures the mind–body contradiction in a paradoxical formulation that plays on themes of being both obedient and in command at the same time. The problem arises from the fact that our mind is confined by its finite earthly origins, the material origins in the physical brain, being a product of complex natural biological processes. Our drive to rise above nature and dominate it is thus restrained by our material nature itself. The concept of objectivity requires that our mind separates itself from matter in order to know reality and command nature from above; yet that is not possible for man to achieve completely. This explains the long tradition of philosophical disputes over the nature of human consciousness, the subjective experience of reality, which further leads to Platonic metaphysics, even questioning the existence of objective reality itself.

When biologists study animals in the lab, such as mice or bacteria, we imagine them distancing themselves, observing their subjects objectively and independently. In contrast, when social scientists study human social behavior their objectivity is challenged by personal psychological biases and cultural norms that interfere with separating the object of study from the subjective interpretation of its observer. Nevertheless, we are driven by biological factors to such a degree that even in studying natural phenomena we are often swayed by the collective mind in our perception of reality. This makes objective scientific work limited by our cultural environment and social construct of reality.

To enhance the power of the individual mind in studying the complexity of our world, we organize into groups. For example our modern academic institutions group people together to advance science, dividing the labor between many researchers in the disciplines that make the social and natural sciences. For a society to engage in studying its own mind, it faces a huge challenge, grappling with a biological social system made of matter and driven by energy flows, working to understand itself. We confront these limitations of our finite brains by working in social groups designed to enhance our capacity, but at the price of limiting the independence of its individual agents. The group dynamics trump the needs of any individual in both our biological and social systems.

A study from the Max Planck Institute for Evolutionary Anthropology characterizes this differential between human cognition and other species as "crucial." Participating with others collaboratively and intentionally necessitates the exclusive rational capabilities of humans, including "especially powerful forms of intention reading and cultural learning, [and] also a unique motivation to share psychological states with others and unique forms of cognitive representation for doing so."[9] The cultural evolution that resulted hundreds of thousands of years ago has equipped mankind with "everything from the creation and use of linguistic symbols to the construction of social norms and individual beliefs to the establishment of social institutions."[10]

In specific support of this cultural intelligence hypothesis, chimpanzees, orangutans, and 2.5-year-old children, were given extensive cognitive tests. The results support the human phenomenon of "shared intentionality" over and against the more simple notion of greater global intelligence. The researchers found that "the children and chimpanzees had very similar cognitive skills for dealing with the physical world but that the children had more sophisticated cognitive skills than either of the ape species for dealing with the social world."[11]

Such shared intentionality is so critical to human functioning that social psychologist Jonathan Haidt deems it "the evolutionary Rubicon" for our species. He offers a concise description of the mechanisms of this "hive mind":

> ...[It is] the ability to share mental representations of tasks that two or more of [our ancestors] were pursuing together. For example, while foraging, one person pulls down a branch while the other plucks the fruit, and they both share the meal...[W]hen early humans began to share intentions, their ability to hunt, gather, raise children, and raid their neighbors increased exponentially. Everyone on the team now had a mental representation of the task, knew that his or her partners shared the same representation, knew when a partner had acted in a way that impeded success or that hogged the spoils, and reacted negatively to such violations.[12]

Our greatest strength stems from the fact that we form cohesive social groups that work together to efficiently divide labor and increase the capabilities and efficiency of the whole. However, this is also the source of our greatest weakness, because for the whole social system to work together in tandem, fully synchronized to maximize its efficiency, it must adhere to certain rules and regulations that limit the operational freedom of its individual components. For this complex system to work, the individual must willingly accept, or be coerced, to be constrained by the requirement and needs of the group.

## Evolution of Morality and Religion

Looking at more particular manifestations of our evolved cooperation, we must consider the species-specific aspects of human morality and religion. In *The Bonobo and the Atheist: In Search of Humanism Among the Primates,* primatologist Frans de Waal asserts that the origin of human morality emerges from within, rooted in our past behavioral adaptations, rising through our evolutionary psychology, rather than being imposed from some authority figure or deity. He argues that our moral behavior is an emergent phenomenon, an evolved psychological adaptation from which religion came about in order to regulate our complex social systems. There is no "inner dualism" that pits man's reason against his instincts, rather, our very instinct has evolved to make us morality-creating beings. Although thinkers like Hobbes and Aldus Huxley and even Freud understood ethics as humanity's "cultural victory" over our depraved nature, De Waal writes, "Everything science has learned in the last few decades argues against the pessimistic view that morality is a thin veneer over a nasty human nature."[13]

Jonathan Haidt discusses how Darwin, from whom generations have inherited the materialist ideas that would seem to belie moral evolution, actually recognized that that "Selfish and contentious people will not cohere, and without coherence nothing can be effected." Darwin supported group selection in *The Descent of Man* by understanding that human character traits that serve little use to individuals, such as sympathy and loyalty, are of definitive use to groups. Haidt goes on to present the image of a superorganism, created by individual bacterium with individualized DNA somehow were encased in one membrane. And like cell mitochondria, wasps and then bees and then ants found themselves in the same situation, "all together in the same hive, they had no choice but to cooperate, because pretty soon they were locked into competition with other hives. And the most cohesive hives won, just as Darwin said."[14]

A half-million years ago, human families also coalesced into tribes, reaping the survival benefits of cooperation. The functional cohesion of survival evolved into a codification of ethical sentiment that created religion. Haidt relies on the metaphor of French sociologist Emile Durkheim that calls humans "homo-duplex," a two-level creature that ascends and descends a mental staircase to move between the mundane and the sacred. With DeWaal as well as evolutionary biologists David S. Wilson and Edward O. Wilson, Haidt sees the religious impulse as an adaptation. "Groups that developed emotionally intense, binding religions were able, in the long run, to outcompete and outlast groups that were not so tightly bound."[15] He goes on to use lofty terms in his Ted Talk to expand on his ideas about this process:

> I don't mean that we evolved to join gigantic organized religions. Those things came along too recently. I mean that we evolved to see sacredness all around us and to join with others into teams and circle around sacred objects, people and ideas. This is why politics is so tribal. Politics is partly profane, it's partly about self-interest, but politics is also about sacredness. It's about joining with others to pursue moral ideas. It's about the eternal struggle between good and evil, and we all believe we're on the good team.
>
> And most importantly, if the staircase is real, it explains the persistent undercurrent of dissatisfaction in modern life. Because human beings are, to some extent, hivish creatures like bees. We're bees. We busted out of the hive during the Enlightenment. We broke down the old institutions and brought liberty to the oppressed. We unleashed Earth-changing creativity and generated vast wealth and comfort.[16]

Furthermore, neurological research bears out religious bonding by indicating that spiritual and religious experiences "activate the brain reward circuits in much the same way as love, sex, gambling, drugs and music."[17]

## Male/Female Dichotomy

As we can see, humans are an extremely social species, building social bonds and extensive social networks that enable the transmission of knowledge and shared intentionality for mutual cooperation. This mutual bonding requires that individuals circle around a common set of ideas, an ideology that forms the basis for shared moral values and religious systems that unite people to work together in harmony to achieve their common goals. However, there are different structures in which social bonding can take shape, from loose bonds that enable a high-degree of independence, on the one hand, to total bondage in a centralized system of control, on the other. In social animals, the biological factors are primarily sex hormones that determine the sexual organization of society and set the disposition for the degree of independence versus dependence among members of the social group. In 1973, sociologist Mark Granovetter proposed a theory on the propagation of information in social networks called "The Strength of Weak Ties," concluding with the paradox that "weak ties, often described as generative of alienation, are here seen as indispensable to individuals' opportunities and to their integration into communities; strong ties, breeding local cohesion, lead to overall fragmentation."[18]

An example of these strong or weak ties can be seen in the masculine versus feminine principles in the social organization of human cultures, which contrast independence and rational self-interest with mutual dependence and altruism—the blind acceptance of the duty to sacrifice for the common good. If the masculine principle of independence is too extreme for anti-social individuals, no social bonding is possible; but if the feminine principle of social cohesion is too extreme then the social ties become too strong, leading to the trap of social bondage. A human society can be considered in hormonal balance if the masculine principle of independence, led by competitive drive and elevated by testosterone, is well balanced with the feminine principle of social bonding, driven by the love and bonding hormone, oxytocin.

In Carl Jung's analytical psychology he identifies the faculty of reason as the Logos and contrasts that with Eros as emotional, feeling, and mythos oriented. This dichotomy, according to Jung, underlies the opposing views of "science versus mysticism," or "reason versus imagination."[19] Jung associates Logos—reason and objective knowledge—with the masculine principle, in contrast to Eros, its female counterpart, expressing inner feelings and intuitions:

> Woman's psychology is founded on the principle of Eros, the great binder and loosener, whereas from ancient times the ruling principle ascribed to man is Logos. The

concept of Eros could be expressed in modern terms as psychic relatedness, and that of Logos as objective interest.[20]

Art Historian Camille Paglia, has contributed immensely to closing the perceived gap between the body and mind and to a historical comprehension of the male-female dichotomy. She articulates ways in which religion manifests in different cultures, favoring the principles of Logos or Eros. In her masterpiece on sexual philosophy, *Sexual Personae: Art and Decadence from Nefertiti to Emily Dickinson*, she illustrates beautifully how testosterone shapes our body and mind throughout the history of Western art and philosophy:

> The book of Genesis is a male declaration of independence from the ancient mother-cults... Mind can never be free of matter. Yet only by mind imagining itself free can culture advance. The mother-cults, by reconciling man to nature, entrapped him in matter. Everything great in western civilization has come from struggle against our origins. Genesis... gave man hope as a man.[21]

Paglia associates the philosophy of the Hebrew Bible with the birth of the Western eye. The Creator of the Universe is immaterial to, or transcendent over the material world, observing it from a dominant posture above it (see Figure 1). This distance from matter enables the omnipotent God to observe objectively, to control and manipulate its shape, as He is not part of matter itself. In Genesis, man, who is made in the image of God, joins the Creator in a dominant position above the created, material world. This dominance posture toward nature is associated with the extreme male brain, a product of a peak testosterone culture. The ancient Hebrews developed an extremely optimistic and benevolent view of existence, and in Genesis 1, God the Father is repeatedly said to have created the world for the good of man.

On the other extreme are the pagan mother-cults, whom Paglia associates with the feminine brain. The mother-cults worship the material world, nature itself, as divine, and hence submit themselves in surrender to its wild and violent, unpredictable and chaotic machinations, which are subjected to the fluctuating natural world of energy and matter. We see this fatalism in the pagan Greek belief in the Moirai (the Fates, see Figure 2), the architects of destiny bestowed upon man that he cannot overcome, as he is trapped in this material world. Plato tried to escape this malevolent world he saw in the flux of matter to a divine realm of perfect ideas that must be totally detached from the reality of human existence on earth. This Gnostic, dualistic worldview separates human existence between the evil and sinful material world and the immaterial world of the soul that is all good.

Figure 1: "God the Father," Cima da Conegliano, c. 1515 [22]

Figure 2: The three Moirai. Relief, grave of Alexander von der Mark (de) by Johann Gottfried Schadow. Old National Gallery, Berlin. [23]

Although the *New York Times* critiques Paglia's conclusions as extreme, they accurately report that she extols the "the spectacular glory of male civilization" and criticizes the narrow scope of feminism, informed by Rousseau's utopianism, that creates "social fiction" by villainizing men. Paglia writes, "Feminism has exceeded its proper mission of seeking political equality for women and has ended by rejecting contingency, that is, human limitation by nature or fate.[24] In her support of the freedom that testosterone-driven Western thinking ushered in, Paglia recognizes the "small-minded" problem of the feminist fiction:

> One of feminism's irritating reflexes is its fashionable disdain for "patriarchal society," to which nothing good is ever attributed. But it is patriarchal society that has freed me as a woman. It is capitalism that has given me the leisure to sit at this desk writing this book. Let us stop

being small-minded about men and freely acknowledge what treasures their obsessiveness has poured into culture.[25]

What she is describing as reflex is a cultivated practice, a story told over and over that has reinforced a perception. Humans, it seems, are quite driven to promote their particular perceptions; and perceptions unchecked can become propaganda.

## Perceptions

It is critical to comprehend how the masculine–feminine dichotomy displays itself in cultures. The swing to one extreme or another creates the very lens through which we see the world. It effects our perceptions of the world and therefore how we organize socially, what gets researched and what doesn't, what kinds of art and entertainment are produced, what journalists report, and what the public latches onto as "truth." In assessing contemporary academic work in the social sciences and humanities, for example, one has to take into account the "spirit of the time"—that is, the academic culture that underpins the social environment in which researchers work and publish. The preface and introduction of this book present numerous examples of how the feminist agenda prevails in Western academia and thereby influences the culture. Feminism is similar to the Platonic dualistic worldview that regards its ideals of an egalitarian, utopian, communal society as products of the mind, not impeded by the limitations of the body. Any mention of biological factors that result in gender and race differences are strictly prohibited in this intellectual atmosphere; in fact, at times they are punished. Jonathan Haidt describes the phenomenon of such insular thinking and how it generates a kind of group trance, in which "moral claims come to feel as objectively true as the claims of physics and math."[26]

These repeating, transformative cycles in regard to our perception of our own mind, human nature, and society, recur throughout history. When social mood is positive, humans feel rising confidence and view themselves as rational beings, above the rest of the animal and natural world. But as social mood declines, humans cannot overcome the increased feelings of irrationality. They come to distrust their own minds and the world around them and become reduced to instinctual, animalistic, group behavior on par with other mammals and even social insects, such as ants and bees, as E. O .Wilson refers to in his work. As we shall further discuss in the coming chapters, all these figures, Charles Darwin, David Hume, E. O. Wilson, and Jonathan Haidt were working in periods of decline in social mood that produced this negative perception of human nature.

During the first half of the nineteenth century, in a period when the patriarchy was in control of academia and the intellect in the West, the Judeo-Christian belief in divine creation by the Creator God, the Father in heaven, came into conflict with Darwin's emerging theory of evolution, arresting its development in the academy. Thus, Darwin had to develop his theory in private, outside of mainstream academic circles. He worked at his own home and was very careful in sharing his ideas, which he only published much later when the culture was more prepared to accept this revolutionary worldview. As I suggest in *The Testosterone Hypothesis*, it was probably a decline in social mood due to decreased testosterone levels that shifted the tides toward the view of Darwinian evolution since the late nineteenth century. This also coincided with the rise of feminism and socialism in our culture. It was not a coincidence that Marx published his work in the same period as Darwin developed his theory. Today, the theory of evolution has gained complete acceptance, and feminism and socialism have utterly taken over the mindset of the academic establishment. There is a concerted effort in the academy to suppress the important discussion of the role of sex hormones as the biological force that shapes our society.

Unfortunately, although the public tends to think of academic science as purely empirical and objective, we are biologically designed to operate under self-imposed and socially induced restraints, even regarding our core belief system and ideology. Our human nature as social animals that form cohesive social groups often prohibits individual scientists from pursuing an independent course of inquiry that contradicts the mainstream line of thinking. This is not different from the kinds of accusations and persecution that can lead to excommunication when heresy is detected in religious groups. For example, the Catholic Church punished Galileo Galilei for publishing his work against its accepted teachings on a geocentric universe.

The view of man as both an independent and a rational being tends to coincide in a culture as a rise in testosterone levels raises our feelings of self-esteem, our internal locus of control, and an optimistic, rational view of human nature. During the period between the 1950s and 2000, there was a rationalist view of man in both psychology and the social sciences, and an individualistic view that regarded man as an autonomous unit, in which the mind of the individual is not part of or overly subject to social group pressure. This view, referred to as *methodological individualism,* also reigned supreme in the study of biological evolution, in which the individual was considered the unit of selection. However, the cycles of history are fractured into sub-cycles of different degrees, and within this overall positive trend in social mood, there were sub-periods of decline, primarily during the recession of the 1970s. During this period of negative social mood, in 1978, social scientist Herbert Simon received a Noble Prize in economics for his work on bounded rationality, his concept that our rationality is limited during decision making, constrained by our finite mind and time resources.

The mainstream view of individual selection in biology, along with reason and atheism versus religion, were represented by evolutionist Richards Dawkins, who published the *Selfish Gene* in 1976. Dawkins presented his thesis that genes are the unit of selection in individuals, in a similar fashion that *memes,* a term he coined for ideas that spread through human society, are the unit of selection in human cultures. The contrast between the overall positive trend in social mood post WWII until 2000 and the negative sub-degree negative trend during the 1970s, is evident in Dawkins' overall approach. While he championed reason, individualism, and science on the one hand, his newly introduced concept of memes, on the other hand, challenged pure reason and free will, because the meme associates human cognition with the biologically determined mechanism of social groups.

During the 1970s some other dissenting voices to the individualistic paradigm emerged, but they were few and usually silenced by the majority. One such scientist was Harvard biologist E.O. Wilson, who published his book *Sociobiology* in 1975, discussing humans as social organisms like ants and other social animals. Another dissenting academic to the individualistic view was the evolutionary biologist David Sloan Wilson who promoted the view of multilevel selection, rather than the individual as the sole unit of selection. This determines that evolution operates on the level of individuals within groups, but also in group selection—groups as a whole system competing against other groups for survival, resources, and propagation of itself.

The societal decline in testosterone and other happy chemicals in our brain since its peak around the year 2000 has resulted in the effect of a decline in social mood. Our collective mind feels depressed and disillusioned with reason, perceiving ourselves too fragile and incompetent to deal with the challenges presented by the times. Jonathan Haidt presented his own model of *social intuitionism* in 2001, in which moral positions and judgments are primarily intuitive, rationalized, justified, or otherwise explained after the fact, and are taken mainly to influence other people to form cohesive coalitions in the social dynamics of group behavior.[27] The idea that the faculty of reason only follows our intuition diverged from the prevailing rationalist model of morality, such as of Lawrence Kohlberg's stage theory of moral reasoning, which was the accepted model in prior decades. We can identify that this trend of attacking human rationality has gained credence in the culture since the year 2000, with another significant psychologist, Daniel Kahneman. Kahneman received the Nobel Prize in Economics in 2002 for his work establishing the cognitive basis for common human errors that arise from people's decision-making processes and biases. He worked in the growing scientific discipline of *behavioral economics* that assesses the emotional influences on choices and how those affect the market. The model that

Kahneman and Tversky published in 1979, called the prospect theory, used cognitive psychology to challenge the previously accepted view in neo-classical economics that rational individuals operate in perfectly efficient markets in which all the information is known to the participants, and their choices are made from such data.

Since the social mood has declined, both individualism and reason have suffered severe blows in our collective consciousness and view of human nature. Currently, the view of man as a rational, independent being is attacked from every side, similarly to how John Locke's rationalist view of man during the eighteenth century Age of Reason was attacked by David Hume and the French romantic philosopher, Rousseau. Locke believed in the *blank slate hypothesis*, which states that the human mind is born with no innate ideas, thus validating the optimistic view of man as a rational being who possesses the absolute power of free will. The opposite view, called *innate ideas*, binds our brains—and our fates—to biological factors, which are associated with the limitations of the material world and animalistic impulses as the basis of behavior.

## Feminism

There are no ideological forces more destructive to the benevolent view of man advocated by the Enlightenment than the unholy trinity of feminism, Marxism, and Eco-feminism. It seems that with the decline in testosterone, which is responsible for creating dominance hierarchies and social order, this triumvirate seeks to demolish the old world order, with the West leading the global hierarchy, by bringing anarchy and chaos to the social system. Established social structures grounded in the biology of the sexes, such as the heterosexual family, gender differentiation, and the sexual division of labor, are attacked in by the progressive vanguard of contemporary academia and media as irrelevant, primitive, and anachronistic, controlled by a religious patriarchy that defies the laws of human progress with outdated traditions. Yet, these powers are blind to the irony that Camille Paglia calls a "delusion" when she writes, "Sexual freedom, sexual liberation—a modern delusion. We are hierarchical animals. Sweep one hierarchy away, and another will take its place, perhaps less palatable than the first.[28]

The prevailing elite ideology of feminism and cultural Marxism views the human personality as a subjective social construct that has no objective biological basis. It is, for example, in complete denial of the role testosterone has in creating the biology of sexual differentiation in the human brain. Moreover, the Marxist-feminists adhere to aspects of Darwinian evolution that attack the biblical view of mankind as a dominant being, created in the image of a transcendent God, above all other animals and above physical nature itself. Hence, by advancing their version of the evolutionary worldview, they seek to reduce the status of man to simply another animal, residing on earth and limited to the material world. In the socialist, utopian, egalitarian version of evolution, evolutionary psychology has no place. The idea that our brains evolved through adaptations to ancestral, environmental conditions clashes with the ideology that there are no innate genders, racial differences, or superior cultures or religions; and this is accomplished particularly by attacking Western culture as having a superiority complex.

Psychological evolution studies our brains also as products of biological evolution—evolved adaptations through sexual selection in human evolution. In 2002 Harvard psychologist Steven Pinker published *The Blank Slate: The Modern Denial of Human Nature*, in which he attacks the model of *tabula rasa* (i.e., blank slate) that prevailed in the mid-twentieth century in the social sciences, making the case that the human brain is primarily molded by adaptations of evolutionary psychology. Nevertheless, Pinker critiques the feminist lockdown on scientific inquiry about sexual difference in the human brain:

> At some point in the history of the modern women's movement, the belief that men and women are psychologically indistinguishable became sacred. The reasons are understandable: Women really had been held back by bogus claims of essential differences. Now

anyone who so much as raises the question of innate sex differences is seen as "not getting it" when it comes to equality between the sexes. The tragedy is that this mentality of taboo needlessly puts a laudable cause on a collision course with the findings of science and the spirit of free inquiry. [29]

This aversion to academic freedom in science is an enormously troubling characteristic of the current culture, and its effects are already broad reaching. Biologist Steve Pruett-Jones writes on the ideological opposition to biological truth:

> One distressing characteristic of the Left, at least as far as science is concerned, is to let our ideology trump scientific data; that is, some of us ignore biological data when it's inimical to our political preferences. This plays out in several ways: the insistence that race doesn't exist, that there are no evolutionarily-based innate (e.g., genetically based) behavioral or psychological differences between ethnic groups, and that there are no such differences, either, between males and females within humans.
>
> But let's look at some data showing prima facie that there are biological differences in behavior between males and females, and that those differences reflect the working of natural selection—in the form of sexual selection—in our ancestors. To do this, we'll use body size as an index of behavior.
>
> It reflects evolved male *behavior*: the tendency of males to compete for females, and the advantage of large body size in that competition. Whether the advantage be in direct competition, so that the larger you are the more you can fight off other males (gorillas, elephant seals), or in female choice, so that females choosing large males can gain protection for her young from marauding males (also gorillas), the difference in size reflects something almost universal among animals: males, who have cheap

gametes, must compete for females who have expensive gametes and invest more in reproduction. And that is why, in study after study in humans, male sexual behavior shows promiscuous mating, while females are more selective.[30]

In failing to publish much new research on the societal decline of testosterone, the highly feminized academic establishment ignores research, for example, that indicates the loss of masculine development among adolescents, who should have been experienced puberty with rising growth hormones levels. Young men between the ages of eighteen and twenty-four in the UK report that even the term "masculine" has negative implications. Only twenty-eight percent of the men polled self-reported as "feeling masculine," and only two percent identified as "completely masculine." This stands in contrast with men in the over-sixty-five generation, fifty-six percent of whom report feeling completely masculine.[31] In concert with their propaganda campaign over the past decades to vilify masculinity in general, feminist academics ignore or suppress research that supports positive aspects of testosterone. A German study indicates the correlation between higher testosterone and a higher incident of honesty in the ninety-one male participants. Co-researcher Dr. Armin Falk suggested that the significantly higher incident of pro-social behavior in those men treated with testosterone might be due to "increased pride and the need to develop a positive self-image in participants."[32] While this outcome is hardly surprising to people who understand the very positive benefits of robust testosterone, what's alarming is that the study results are reported as surprising and contrary to some long-held canon regarding the anti-social behavior of higher testosterone—as if that is a truth that certain newer studies are only now refuting. In fact what this depicts is the bias inherent for decades in most research concerning the hormone; and that is a direct result of the encroaching feminist-Marxist ideology that has come to grip the research community.[33]

This vice grip on the academy and the power of the feminist lobby was vividly portrayed at the highest levels of academia when, in

2005, Harvard University President Larry Summers was forced to resign. He had spoken at a conference, citing that biological research might offer insights as to why men continued to predominate on mathematics, engineering, and hard-science faculties. The very notion that Summers referred to innate male-female differences sent the cultural elite into a frenzy; and supported by the media, the liberal shock troops managed to oust one of the most powerful people in higher education. This event dealt a fatal blow to any further academic discussion on the subject of innate sex differences. While some reported on these events as mere "careerism" on the part of feminist academics trying to procure more posts for female colleagues, it has to be understood as much more pernicious.[34] The ideologues demanding "tolerance" are so utterly intolerant of dissent from their canon that no one is safe from their purgative methods—not even the president of Harvard. In this atmosphere it becomes all the more clear why thoughtful researchers and writers present theories with gaping holes or contort themselves to avoid potential controversy, because this kind of controversy might actually end their careers. This would explain why social scientist Jonathan Haidt never discusses gender differences in his moral foundations theory. He presents liberals (who have feminine brains) only as concerned with empathy and compassion, disregarding the impact of testosterone and gender proclivities in the discussion.

Even the most obvious biological facts about male–female differentiation are taboo to highlight or possibly make policy decisions on, such as men's stronger bones, muscles, and ligaments. Men have forty percent more muscle mass than women, and this has significant ramifications in military policy, for instance, regarding women in combat.[35] Nevertheless, the feminist lobby insists on pushing the US Congress and military to equalize opportunity for women to fight alongside men in all forms of warfare, regardless of the greatly increased risks to women—all for the sake of a skewed idea of "equality."[36]

The ideological emphasis on equality, whether in feminist or socialist thought and policy, is a notable result of declined testosterone in the culture. Empathy is the feminine characteristic of motherly love and compassion, and the wave of empathy currently sweeping the culture compels people with the ideals of fairness, non-judgment, and every manner of equality. *New York Times* reporter David Brooks cited the many books being written about empathy in 2011 when he wrote an op-ed titled *The Limits of Empathy*. He quotes Pinker as characterizing the phenomenon as "an empathy craze;" however, Brooks explains, research does not bear out that the *feeling* of empathy has any effect on actual moral motivation or action. He goes on to describe how empathy is used as a smokescreen for actual action:

> These days empathy has become a shortcut. It has become a way to experience all the delicious moral emotions without confronting the weaknesses in our nature that prevent us from actually acting upon them. It has become a way to experience the illusion of moral progress without having to do the nasty work of making moral judgments. In a culture that is inarticulate about moral categories and touchy about giving offence, teaching empathy is a safe way for schools and other institutions to seem virtuous without risking controversy or hurting anybody's feelings.[37]

A *Wall Street Journal* writer Paul Bloom, taking Brook's concern one step farther, describes empathy as downright perilous. Especially when it comes to political policy, Bloom's concern is that "[e]mpathy distorts our reasoning and makes us biased, tribal and often cruel."[38] He lays out a number of studies in which "empathy-driven judgments" result in assistance for one person but not for multitudes. The nature of such emotionally informed inclinations set people up to be manipulated and to disregard "practical calculations." According to Bloom, what happens in reality, separate from sad and compelling video or news stories, is a "sort of perverse moral mathematics... It's why people's desire to help abused dogs or oil-drenched penguins can often exceed their interest in alleviating the suffering of millions of people in other countries or minorities in their own country."[39] Empathy out of balance with rationality can be downright foolhardy.

# Sex Wars

Feminism of the later twentieth century cannot be understood separate from another pillar of utopianism, Marxism. Cultural Marxism is an ideology that evolved from the nineteenth and twentieth century political ideology of communism. According to Marx, capitalists exploit the working class; but as appropriated by feminists, that exploitation extends to all men—the exploiting patriarchy that determines the social dominance hierarchy. Male members of society are on top and oppress the women at the bottom of the social ladder. The premise of oppression bleeds into any scenario wherein one, more privileged group suppresses another: economic, racial, sexual identity...the variations of the victim paradigm are seemingly innumerable. In the environmental realm, we find the Eco-feminists, who view men as exploiters of Mother Earth through their drive for dominance over nature. Such dominance can be exercised through science, business, and industry—any endeavor that necessarily manipulates nature as an object for human consumption and enjoyment. Eco-feminist Marxists are in all-out war against the West as a force of white-male cultural dominance over indigenous societies in other parts of the world, as well, such as Africa, the Middle East, and South America. According to this worldview, other societies have only suffered under the Western capitalist system of exploitation. The feminist–socialist axis is in a power struggle to destroy Western men and give no quarter to the manifold good those men and their ancestors might have accomplished throughout the centuries. Once again, Camille Paglia is not afraid to call her feminist sisters out:

> Men have sacrificed and crippled themselves physically and emotionally to feed, house, and protect women and children. None of their pain or achievement is registered in feminist rhetoric, which portrays men as oppressive and callous exploiters.[40]

Eco-feminism is a term coined in 1974 by the French writer Françoise d'Eaubonneto to link feminism to ecology. Radical ecologists and feminists seek to assert the feminine instinct to nurture Mother Nature by fighting against the masculine motive of domination and the capitalist exploitation of the earth and its resources. The feminist ethos of liberation from male dominance has been projected as the struggle of feminine nature against the masculine culture. Man's split from nature was his Original Sin that can only be healed by holistic union with the whole of nature and by ending oppression of the feminine.

In *The Death of Nature: Women, Ecology and the Scientific Revolution*, published in 1980, prominent Eco-feminist Carolyn Merchant writes that the male brain sought to objectify and atomize nature as inert matter:

> The female earth was central to organic cosmology that was undermined by the Scientific Revolution and the rise of a market-oriented culture...for sixteenth-century Europeans the root metaphor binding together the self, society and the cosmos was that of an organism...organismic theory emphasized interdependence among the parts of the human body, subordination of individual to communal purposes in family, community, and state, and vital life permeate the cosmos to the lowliest stone.[41]

Like many utopians, Merchant claims a more pure and idyllic era—in this case the pre-Enlightenment 1500s—without drawing attention to the difficulties or drawbacks of the era. She anachronistically criticizes Francis Bacon's usage of a feminine metaphor to depict the exploitation of nature. Bacon stated that nature should be "bound into service" and enslaved unto human ends to regain our dominion over it, which was lost in the fall from grace in the Garden of Eden:

> [S]he is either free,...or driven out of her ordinary course by the perverseness, insolence and forwardness of matter and violence of impediments...or she is put in constraint, molded and made as it were new by art and the hand of man; as in things artificial...nature takes orders from man and works under his authority.[42]

Agreeing with the commanding language of Bacon, Paglia maintains that the very remaking of reality is the way of nature. Further, she states, "Most of Western culture is a distortion of reality. But reality should be distorted; that is, imaginatively amended."[43] To not amend reality is to do an injustice to human potential.

Hardly confined to the ivory towers of academia, Eco-feminist thought is readily found in popular media. A 2016 *Washington Post* article leads with "Your manliness could be hurting the planet" and goes on to present research about the "gender eco-friendliness gap."[44] Not surprisingly, women outdo men in their green living and consuming. A researcher named James Wilke and his cohort theorize that men are driven to bypass or oppose environmentalism "in order to safeguard their gender identity."[45]

## Transgenderism

As the triune influence of feminism, Marxism, and Eco-feminism have saturated the thought leaders of our society, perhaps no one consequence is more stunning than the swiftly growing phenomenon of gender identity politics in all manner of "transgenderism." Perception and politics have driven science and contorted public opinion into not only tolerance or acceptance of gender variation, but enthusiastic promotion thereof. Professor Paglia is very straightforward in criticizing the current "transgender mania" and in recognizing it as a characteristic of a culture's late phases.[46] Bringing her broad historical knowledge to bear, she explains that she has seen this cycle before; and what she describes is clearly the phase of cultural decline caused by falling testosterone levels:

> I found in my study that history is cyclic, and everywhere in the world you find this pattern in ancient times: that as a culture begins to decline, you have an efflorescence of transgender phenomena. That is a symptom of cultural collapse.[47]

Moreover, Paglia calls the "wildly inflated claims" of transgenderism propaganda.[48] The lies promoted to reinforce this mania include the anti-biological idea that sexual reassignment surgery and therapy can actually change the cellular DNA coded with a person's sex. She goes so far as to label it a "kind of child abuse" when misinformed parents, school officials, and health care workers encourage sex change in young people.[49] That child abuse becomes all the more widespread when government entities, the media, and entertainment outlets propagate the mythology of multiple gender identities and help create a pervasive culture that makes it almost impossible for families to counteract what their children are being taught is "normal."

At the university level, such "social constructionism" is disputed by outspoken critic Dr. Jordan Peterson of the University of Toronto. Peterson is a professor of psychology, who has become controversial in his opposition to cultural Marxism and the transgender culture. Like Paglia, he claims that students are being taught lies as legislation is being generated that codifies the subjectivity of personal identity:

> They're detached from the underlying biology and from the underlying objective world. So Bill C-16 contains an assault on biology and an implicit assault on the idea of objective reality. It's also blatant in the Ontario Human Rights Commission policies and the Ontario Human Rights Act. It says identity is nothing but subjective. So a person can be male one day and female the next, or male one hour and female the next.[50]

Whether in the US, Canada, or Thailand, where legally acknowledging a third gender is under consideration,[51] Peterson is correct when he calls the situation "bleak."[52] A 2016 study reveals that thirty percent of self-identifying transgendered youths have tried at least one time to commit suicide. Compare that with the overall national average of teen suicide attempts at around eight percent. There are similar differentials in incidents of self-harm, like cutting, and these numbers do not change significantly following sex-reassignment treatments.[53] While the transgender narrative maintains the reasons for these high rates of self-destructive behavior are found in societal pressures and intolerance, the academic and media elites will not even consider how hormonal deficiencies or imbalances might create not only the gender dysphoria but also the underlying depression.

As this chapter has sought to present, this is the tragic outcome to the terribly irony we find ourselves in: the condition of diminished testosterone creates the societal mood and perceptions that resist a comprehensive investigation into conditions caused by the diminished testosterone. The more persuaded people are by the propaganda, the more intractable they are, unwilling and unable to see objective truth. The clarion call in the wake of the holocaust was "Never forget," but Dr. Peterson makes the important point that if we don't understand a situation, we cannot remember—or rightly remember—what led to it or what exacerbated it, and therefore we are stymied in how to cure or prevent it.[54] In this book's efforts to shed light on comprehending our perilous times and offer reason for hope, chapter 2 will delve deeper into explaining human dominance hierarchies in both their endocrine foundations and their societal outworking.

# 2

---

# DOMINANCE AND SUBMISSION

*Sexual freedom, sexual liberation. A modern delusion.*
*We are hierarchical animals. Sweep one hierarchy away,*
*and another will take its place, perhaps less palatable*
*than the first.*[1]
— Camille Paglia

*Map the masculine on to the dominance hierarchy. Map*
*the feminine on to nature. Nature is that which selects.*
*Women select, they are not like chimps, they are selective*
*maters. They are nature.*[2]
— Jordan Peterson

Testosterone, as well as being a sex and growth hormone, regulating our reproductive life cycle, is also a dominance hormone, structuring our social hierarchies. Pursuing a dominance position in a social hierarchy requires potency and strength to compete with rivals. Therefore, being dominant, sexy, healthy, and strong are all part of the same physical and social identity.

To understand why dominance hierarchies emerged in the evolution of social animals we begin by exploring the origins and purpose of sex. Evolutionary psychology determines that natural selection in human behavior provided us with psychological adaptations to specialize in the environment in which we evolved.

# Sex Wars

Sexual selection leads an organism to adaptations in mating opportunities that enhance its sexual reproduction. Male mammals have a high rate of potential reproduction given their enormous quantity of sperm production; therefore, sexual selection provides a high sex drive as a behavioral adaptation that makes males more determined to compete for sexual access to females. Female mammals, by contrast, produce few eggs relative to male sperm production in the course of their fertile life cycle, thus having low reproductive capacity. As a result, choosiness is an evolved adaptation to select the better quality mates. In short, because eggs are expensive and sperm are cheap, the males compete for the females, who show off their value as fertile mates so that the males compete for them. Hence, males seek social status and dominance to secure their reproductive activity through competition. The male has long been considered the competitive gender, while females seek security and are associated with less risk-taking behavior. Later in this chapter, we will look at a contravening hypothesis to this model, but those ideas in no way contradict that males compete for females' acceptance of them as mates.

Not surprisingly, research with humans finds that female attractiveness is valuable to men in choosing mates. In particular, physical symmetry in the woman's body is a fitness indicator that men prefer in study after study. Evolutionary psychologist Geoffrey Miller, in his book *The Mating Mind*, claims that sexual selection is even more important than natural selection, particularly in how our complex minds evolved. Like the ornately spectacular tail of a peacock that evolved only for the purposes of courtship, Miller sees the ornate capabilities of man's mind as foundationally "courtship centered," which subsequently allows us to be survival oriented.[3] A *Newsweek* article explains the preference for more attractive females succinctly:

> Studies from around the world have found that while both sexes value appearance, men place more stock in it than women. And if there are social reasons for that imbalance, there are also biological ones. Just about any male over 14 can produce sperm, but a woman's ability to

bear children depends on her age and hormone levels. Female fertility declines by two thirds between the ages of 20 and 44, and it's spent by 54. So while both sexes may eyeball potential partners, says Donald Symons, an anthropologist at the University of California in Santa Barbara, "a larger proportion of a woman's mate value can be detected from visual cues." Mounting evidence suggests there is no better cue than the relative contours of her waist and hips.[4]

Human evolution has also created the sexual division of labor in order to facilitate greater efficiency and productivity in resource production and utilization. We can observe that females are more inclined toward empathy and are more interested in forming social bonds, as they are wired to rear and care for their children. Males are more interested in physical objects and understanding their environment in order to be productive and provide the sustenance for their family. The empathizing–systemizing (E–S) theory, originated by Professor Simon Baron-Cohen, professor of Psychology and Cognitive Neuroscience at the University of Cambridge and the director of university's Autism Research Center, suggests that this is the reason that men are overrepresented in engineering professions, while women prefer social work and educating children.[5]

**Mutation, Variation, and Sexual Recombination**

Evolution must produce variation in order to allow for selection to take place for the survival and propagation of the fittest. Genetic mutations cause variations in the genome, but to further spread them in the population of the species, sex is necessary to mix different gene sets and to protect the genome of the species from harmful mutations that cause maladaptation, disease, and death. The mixing of genes in sexual reproduction is the method that emerged in evolution to allow for the recombination of genes.

Human DNA comprises twenty-three pairs of chromosomes. Asexual reproduction, called *mitosis*, in which an organism simply clones its genome, requires only an egg and no sperm to replicate itself and split in two. With mitosis, the offspring has the same genes as its parent. Sexual reproduction, on the other hand, involves *meiosis*, a process in which the cells' DNA doubles, its genes are shuffled, and it divides into four cells. Each cell contains only half the DNA from the parent cell. To form a new organism, the egg and sperm fuse together to further mix their genes and diversify their genetic heritage.

In recent years there has been a paradigm shift in the scientific understanding of the purpose of sex in evolution.[6] [7] The previous, century-old paradigm determined that sex increases genetic variation by producing variety in the genome from which the best adaptations prevail in subsequent generations. Even if there are costs to meiosis (e.g., it takes two parents to create just one offspring), the reigning model focused on the advantages of variety. But what if the very fact of mutative variety is the "problem" and sexual selection is the "solution?" Newer thinking in the field posits that sex is the process of filtering disadvantageous mutations. It is the mechanism that limits the proliferation of deleterious mutations:

> [E]vidence for sex decreasing genetic variation appears in ecology, paleontology, population genetics, and cancer biology. The common thread among many of these disciplines is that sex acts like a coarse filter, weeding out major changes, such as chromosomal rearrangements (that are almost always deleterious), but letting minor variation, such as changes at the nucleotide or gene level (that are often neutral), flow through the sexual sieve. Sex acts as a constraint on genomic and epigenetic variation, thereby limiting adaptive evolution.[8]

Whether the purpose of sex is to spread genetic variation through the gene pool of the species, to maintain the integrity of the gene pool in a large population, or to reduce deleterious mutations, it is clear that sex is used to reshuffle the genes in given population of a species. Understanding the purpose of sex in reshuffling the genes is crucial to realizing why both biological evolution and cultural evolution require sex wars to mix the genes of different organisms and groups. In other words, why males must compete for females to choose them and further propagate their DNA. Now that we have introduced the purpose of sex in reproduction, we can proceed to study how evolution creates dominance hierarchies in structuring the sexual reproduction of social animals.

## Dominance Hierarchy

In many mammalian social groups, the males establish a dominance hierarchy. This arises in social groups as a result of the interactions of the members, often through aggression, to establish a system of ranking order. Inside the social system of the group, individuals compete for dominance over limited resources, such as valued territory, food, and mating opportunities. This creates an efficient, orderly system in which relative relationships are formed according to rank between same-sex individuals, preventing the group members from fighting each other all the time. Repetitive interactions lead to the establishment of the prevailing social order in this dynamic complex system, which is subject to change by subordinate animals that are motivated to challenge the authority of superiorly ranking members.

In the evolution of sex and reproductive strategies, various systems of social orders have emerged. Social animals tend to arrange themselves on a continuum from an egalitarian to a despotic society. In the former, members exist with equal ranking; in the latter, all members align under one, dominating authority. Between these extremes, a hierarchical system gives members rankings that create a "linear distribution of power," in which dominance or submission is relative to other group members. An example of an egalitarian system in nature is how a cackle of hyenas function, while bands of gorillas maintain a despotic system.

The major benefits of dominance are foraging and reproductive success for the top ranking members. This is evident in primates, and particularly in dominant male baboons, which have more access to females. Offspring of female baboons that are higher ranking enjoy higher rates of survival. However, scientists have discovered that there costs for such high dominance, mainly higher metabolic activity and increased stress hormone levels.[9] [10]

More recent research indicates that the stress hormones, such as cortisol, that also play a part in structuring dominance hierarchies, are higher in the dominant alpha male, in contrast to what has been the previously accepted theory:

> In social animals, reproductive success is often related to social dominance. In cooperatively breeding birds and mammals, reproductive rates are usually lower for social subordinates than for dominants, and it is common for reproduction in subordinates to be completely suppressed. Early research with captive animals showed that losing fights can increase glucocorticoid (GC) secretion, a general response to stress. Because GCs can suppress reproduction, it has been widely argued that chronic stress might underlie reproductive suppression of social subordinates in cooperative breeders. Contradicting this hypothesis, recent studies of cooperative breeders in the wild show that dominant individuals have elevated GCs more often than do subordinates.[11]

Research shows that stress hormones rise in the dominant members of baboon society in relation to lower ranking individuals, particularly in times of conflict when the hierarchy is in transition. It appears that the individuals challenged from below their rank experience greater stress levels in the transition phase. Basal cortisol levels are higher when more dominant creatures are "rank shifting," even if the system at large is relatively stable. Interestingly, this rise in cortisol comes when males within just a few steps of rank are challenging the more dominant male—when there is an imminent threat of challenge. Yet, when that same male is making the challenge to move up in rank, cortisol levels are not elevated.[12]

So it is that the downside to alpha male status comes in defending their mating partners and their territory, which requires a lot of energy and resources and leads to a greater loss of body mass, a decline in physical health over the long haul, and thus rapid aging. Alphas' reduced fitness results in a high rate of replacement after short periods as a result of physical strain. Because a high rank includes both benefits and costs, individuals will weigh the opportunity to gain rank given their characteristics of age and physical strength. Furthermore, all members of a group do better in a stable hierarchical environment. Less conflict that could upset the hierarchy leads to fewer injuries and ultimately higher fertility rates. To sustain such social order, members need to understand and adhere to their respective roles in the pecking order. Given the advantages of less stress, better health, and longer lives for the non-dominant or beta males, the competition against alpha males is mitigated. Subordination has its advantages that often outweigh the disadvantages.[13] Conversely, however, subordinate members suffer from less access to reproductive activity and to food, as they are often last to feed. They also have less access to shelter and nesting sites.

Nevertheless, stable hierarchical environments seem to be maintained by the dynamics that researchers call the *interpersonal complementarity hypothesis*. This hypothesis recognizes that individual social behavior is met with relatively predictable corresponding action from its counterpart. This thesis suggests complementary behaviors of authority and obedience, as friendly gestures will be met with such, and hostility will raise greater enmity.

It is clear that the dominance hierarchies in social animal groups are fundamentally fueled by hormones. Research with wasps, for example, and naked mole-rats show that the dominant female suppresses the hormonal activity of both male and female subordinates, thereby guaranteeing access to reproduction.[14] While it was earlier hypothesized that a queen's pheromone secretion suppressed the sub-dominants' reproductive hormones, more recently scientists recognize that these dominant females have very high levels of circulating testosterone:

> In sub-dominant males, it appears that luteinizing hormone and testosterone are suppressed while in

females it appears that the suppression involves the entire suppression of the ovarian cycle. This suppression reduces sexual virility and behavior and thus redirects the sub-dominant's behavior into helping the queen with her offspring.[15]

Effectively, the queen "bullies" the others into submission through "intense dominance and aggressiveness on the colony."[16] This comprehension of testosterone's activity in the lead female of mole-rat society can aid our understanding of human behavior around sexual selection, as well. It allows for a different lens through which to see human competition for reproduction.

Although female dominance is observed in only a few mammals, such as hyenas, lemurs, bonobos and elephants, when adult males act submissively during feeding and grooming, newer research concludes female mammals nonetheless do compete among each other for reproductive dominance. This is evidenced in degrees of ornamentation, as with birds' plumage and primates' perineal swelling and coloration, as well as the evolution of "weaponry" (e.g., nails, teeth) in females. The significance in this broader understanding of sexual selection is that it overturns a long-held bias to stereotype males and females with extremes of opposing characteristics in mating. Males are aggressive, females passive; males are competitive, females picky. In a Cambridge University project, researchers assert that the practices and evolution of sexual selection processes should be understood qualitatively instead of quantitatively. In other words, both sexes exhibit competition among themselves in addition to how they act across gender lines. Therefore, it is important to study them as separate behavior systems that are intertwined.[17] Moreover, it is important to recognize the significance of the endocrine system at these essential levels of societal structuring.

## Dominance in the Food Chain

Having reviewed the function of dominance hierarchies in the animal kingdom, we will turn our attention to the human realm. In that connection, it is worth noting that the human species has dominance over the animal food chain as the top predator. In fact, the evolution of our rational abilities as Homo sapiens might be particularly linked to the fact that our ancestors not only ate meat, but that they cooked their meat. The dense calories of meat powered our brains to grow at a faster rate and to grow more complex than vegetarian gorillas, for example. Evolving to cook food gave us more options for caloric intake, because cooking acts to preserve meat and also releases certain nutrients, making them more bioavailable to our bodies. Meat was readily available in the African areas where we most likely evolved and simply is "the best package of calories, proteins, fats and vitamins B12 and B9 needed for brain growth and maintenance."[18]

Our dominance over the natural world, the ecosystem, and the lower animals is fueled by healthy levels of the dominance hormone testosterone. It aligns with the biblical command from God the Father in Genesis 1: "God blessed them and said to them, 'Be fruitful and increase in number; fill the earth and subdue it. Rule over the fish in the sea and the birds in the sky and over every living creature that moves on the ground'" (NIV). When we see in history an aversion to our biological imperative to dominate the food chain and freely partake of the dense proteins that fuel us, it is likely an era of testosterone collapse. Declining testosterone leads to the vegetarian, Eco-feminist propaganda that eschews a high-fat, high-cholesterol diet, because it characterizes humanity's natural state as one of myth:

> Claiming to be at the top of the food chain has become a popular justification for eating animal products and an affirmation of our ability to violently dominate everything and everyone. Yet justifications for needless violence that draw on notions of power and supremacy are based on the philosophy of "Might makes right" — the principle behind the worst atrocities and crimes of human history.[19]

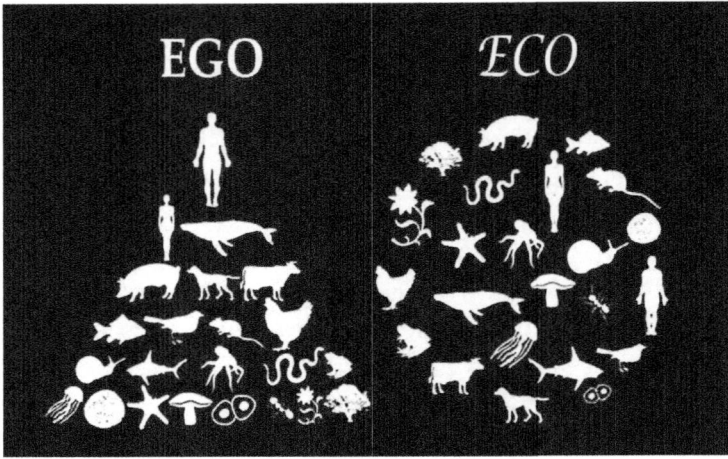

Figure 3: Ego vs. Eco (Original image altered per Creative Commons license. "Ego" and "Eco" replace original heading) [20]

Figure 3 depicts the philosophy of Eco-feminism quite succinctly. When humans live in their understanding of "harmony" with nature, the perfect circle of life is maintained. When mankind dominates the natural order, we are guilty the evil of "ego" and the destruction of our surrounding ecosystems. As described in this book's introduction, the Eco-feminist movement is a pagan cult that bows down in submission to Mother Nature. It has been resurrected in various forms throughout human history, but none so infamous as the Eco-feminist cult of nature worshipers and vegetarians that were Hitler and the Nazis.

# 2 • DOMINANCE AND SUBMISSION

## The Dominance Hierarchy in Human Societies

Figure 4: March of Progress [21]

The *March of Progress* is a well-known illustration that presents a visual schematic of the evolutionary stages leading to Homo sapiens over twenty-five million years. The primary discernible progress in the modern human species is the upright posture and tall form of bipedal walk. The rise of man to an erect posture and dominance over nature from his primitive ancestry in apes is a remarkable progress of evolution. However, man still shares much of his neurobiology with his mammalian heritage, and by studying this linkage we can deduce our own behavioral biological impulses.

The Darwinian evolutionary struggle for the survival of the fittest is a constant battle of organisms over resources, such as food, territory and the biological drive to "be fruitful and multiply," or reproduction. Because resources are limited, organisms often compete violently and aggressively over their availability. Coalition building and group organization are among the strategies that have evolved in primates and mammals for the propagation of the species. The drive for dominance over resources includes gaining territory for the group as a whole, as well as status and position among the group members, in multilevel group selection. In social animals, such as chimpanzees, as well as human, we observe the multilevel struggle for dominance and status in both the intergroup settings and inside the group level. Serotonin, the status and pride hormone developed very early in evolution, and it governs dominance hierarchy in ancient animals, such as invertebrates, including lobsters. Testosterone, the dominance hormone, exists only in vertebrates and regulates dominance behavior along with serotonin.[22]

It is been discovered that our aggressive social behavior most likely originates in our primate evolution, as evidence of warfare in our closest "ancestors," chimpanzees, has been observed. Full-scale warfare with one group fighting another or a group attacking an individual is not seen often among animals. Such "coalitionary violence" exists with ants and some other social insects, and certain mammal groups like wolves, hyenas, and lions.[23] Nevertheless, to see about our immediate evolutionary heritage, we turn to the leader in chimp research, Dr. Jane Goodall. Dr. Goodall started observing a group of chimpanzees in Tanzania's Gombe National Park in the early 1960s. For a long period she viewed chimpanzees as rather non-aggressive in their behavior, until surprising events developed in the 1970s. The group of chimpanzees started to split into two different groups, and later one group opened a gradual war of annihilation against members of the other. The Gombe Chimpanzee War occurred after a split in 1971, followed by a four year war from 1974 to 1978.[24]

As has been shown in this book, research demonstrates that testosterone drives masculine self-confidence and strength, as well as an internal locus of control and one's self-interest before other members in the group or altruism toward other groups. This is rooted in testosterone motivating one's own survival, interests, and reproductive impulses—the self over the other. Hence, it explains why testosterone is at the root of sociopolitical conflict over both the sexual and political organization of our society. A study done in 2013 by Danish scientists found that stronger men correlate with more conservative political and economic views. The study was conducted among hundreds of men in the US, Denmark, and Argentina, and the researchers concluded that men's relative physical strength is indicative of their ability to defend their families and procure resources to provide for them:

> This is among the first studies to show that political views may be rational in another sense, in that they're designed by natural selection to function in the conditions recurrent over human evolutionary history.[25]

In addition to changes in testosterone that affect social mood, a less obvious hormone that might play into the dynamic of group conflict is oxytocin. Excesses in the more "feminine" bonding hormone that fosters mother–child bonding during nursing, for example, can lead to a lack of masculine independence and, in turn, promote ethnocentrism or racism. When dopamine levels rise and in-group bonding increases in one direction, sensitivity about perceived threats from out-groups can increase alienation. In 2011, Professor Carsten de Dreu at the University of Amsterdam conducted research that indicated this inverse relationship between the "cuddle hormone" and tolerance toward those outside one's group.[26] In each of his studies, subjects in groups who were give oxytocin showed much increased favoritism to their own group and either no change in perception of the out-group or increased negativity.[27]

Considering the evolution of intergroup warring in humans, Dutch scientist Mark van Vugt has developed the Male Warrior Hypothesis (MWH) that contends such conflicts have been a critical part of human psychological development. Succeeding at between-group competition and conflict expands a society's terrain, resources, and mating options, as well as status. Therefore, males, in particular, were compelled to evolve in-group preferences and out-group aversion, such as the drive to instigate aggression against outsiders. Female aversion to males from other groups came from their heightened risk of rape.[28] So universal is the in/out-group bias that researchers can create it in studies over something as benign as which picture study participants prefer. Van Vugt's hypothesis accounts for the fact that men tend to remain in the group they were born to, whereas women would have gone to the group their mate originated from, thus men developing higher out-group aversion than women, which studies indicate. Men are more debasing of others they see as outsiders, they measure higher in loyalty toward their own group than women do, and they are more inclined to cooperate with others when their group is threatened.[29] Intergroup, and specifically male on male, violence, evolved as a powerful facet of sexual selection and persists in modern mankind.

## Culture of Submission

Testosterone collapse in the West leads to the election of leaders on the anti-testosterone side of the pendulum, such as President Barack Obama and the Catholic Pope Francis. Both play on the feelings of denial and annihilation of the self and the masculine ego, while promoting the feminine social values of empathy, altruism, and sacrifice. Pope Francis comes out of a strong socialist strain in Argentina, and President Obama was notably influenced by communist sympathizers throughout his formative years. Obama became known for bowing in submission to world leaders who were anti-Western. After President Obama was criticized in April 2009 for bowing to King Abdullah of Saudi Arabia, in November 2009, the President was again accused of being "treasonous" when he bowed to Japanese emperor Akihito. This issue at hand is not displays of respect, but the longstanding American practice of not deferring to royals or dictators in their protocols of subservience. In the case of his bow to the Saudi king, *The Washington Times* wrote, "By bending over to show greater respect to Islam, the US president belittled the power and independence of the United States.[30]

In the peak testosterone philosophy of the Hebrew Bible, we see the dominance effect in the social organization of religious groups. Exodus 23:24 states, "Thou shalt not bow down to their gods, nor serve them, nor do after their works: but thou shalt utterly overthrow them, and quite break down their images." In Leviticus 26:1, the Israelites are exhorted, "Ye shall make you no idols nor graven image, neither rear you up a standing image, neither shall ye set up any image of stone in your land, to bow down unto it: for I am the LORD your God."

This issue of bowing, of our physical posture in relation to others is not insignificant. Not only is it an outward gesture of an inward mindset, it carries both physiological and sociological ramifications. Physically, bowing, kneeling, and prostrating the body brings down the heart rate and changes the activity of the endocrine system. Socially, they are universal signs of submission—and not just with humans. Submissive chimps act in all these same ways when encountering a dominant chimp, as do canines and other animals within their own groups. These postures communicate a lack of threat and obeisance to the superior.[31] Harvard researcher Amy Cuddy has found that to increase testosterone and lower cortisol, a person need only modify his or her body language. The more erect and expansive, the more confident and assertive one feels.[32]

Cuddy's work has been challenged recently by critics who claim her findings failed to be replicated in other studies under similar conditions, but Cuddy asserts her confidence in her science:

> I have confidence in the effects of expansive postures on people's feelings of power — and that feeling powerful is a critical psychological variable. As Columbia University Professor Adam Galinsky and colleagues wrote in their 2016 review, a person's "sense of power…produces a range of cognitive, behavioral, and physiological consequences," including improved executive functioning, general optimism, creativity, authenticity, the ability to self-regulate, and performance in various domains, to name a handful.[33]

Furthermore, according to the Testosterone Hypothesis, it is actually hormones that drive dominance behavior, not the other way around, so Cuddy's error might be the result of having reversed the causal link between hormones and behavior, not because there is a lack of causality.

# Sex Wars

*Submission in Islam*

The Muslim religion capitalizes on the actions and attitudes of a strongly delineated hierarchy, including the physiological factors just discussed, as seen, among other ways, in the prayer posture of its adherents. Five times a day the Muslim bows his face to the earth in repetitive cycles of prayer. More than simply individual piety is on display in this ritual. To know that all Muslims throughout the world are pledging their fidelity and obedience to Allah, the prophet Mohammed, and their worldly representatives, is profound and powerful in binding the faithful together. Islam, in fact, means submission. The difference between the East and West in Eurasia is visible in the posture of prayer. In the Jewish tradition of the West, people pray in an upright posture, looking above to the Creator in the heavens. A common phrase in Judaism regarding the worship posture toward Yahweh is to "spread out" or "stretch out" one's hands:

> And Moses said unto him, As soon as I am gone out of the city, I will spread abroad my hands unto the Lord; and the thunder shall cease, neither shall there be any more hail; that thou mayest know how that the earth is the Lord's. . . . And Moses went out of the city from Pharaoh, and spread abroad his hands unto the Lord: and the thunders and hail ceased, and the rain was not poured upon the earth. (Exodus 9:29, 33)
>
> It came to pass, when Moses held up his hand, that Israel prevailed: and when he let down his hand, Amalek prevailed. (Exodus 17:11)

Likewise, in various Christian traditions, the raising of hands is prevalent as a position of worship, following the Jewish practice and such New Testament instruction as Paul's writing to his protégé Timothy: "I will therefore that men pray everywhere, lifting up holy hands, without anger and disputing" (1 Timothy 2:8).

By contrast, the low-testosterone, feminine, Asian mystical religions, such as Buddhism and Islam, bow down in submission to pray toward the earth, in the manner a mother-earth cult worships nature. In the case of Islam, however, the religious faithful have been trained for centuries to obey the figure that one writer calls "history's supreme alpha male."[34] Bill Warner writes about submission in his work on political Islam in the article *The Two Kinds of Dhimmis:*

> Dhimmis begin with Mohammed. He was the world's supreme master of making others submit to his will. Mohammed had the insight into the human psyche that all human beings have a genetic disposition to submit to the will of the group and higher ranked individuals.
>
> Previous religious leaders and philosophers approached humanity with the idea of freeing the individual from fear. Mohammed did not try to free humanity, but to make humanity a slave to Allah, the god of fear. So he "revealed" the ultimate alpha—Allah. Under Allah, all humans come to their fulfillment by being Allah's slave. But since Mohammed was the only "prophet" of Allah, to obey Allah was to obey Mohammed. Islam is submission to Allah/Mohammed.[35]

The dynamic of the Muslim dominance hierarchy is also displayed in how young men are manipulated into jihad activity of the most treacherous kind. This manifestation of sex wars comes when young Muslim men are restricted from any sexual outlet—sometimes well into their thirties—and the prospect of sexual satisfaction in the afterlife is presented them. Because they are expected to remain chaste until marriage, and marriage is not quickly forthcoming for various reasons, "recruiters lure potential suicide bombers with assurances that blowing themselves up will give them perpetual access in the afterlife to 72 beautiful virgins."[36] Dr. Tawfik Hamid, an Islamic scholar, writes about the abuses of sex and power that compel young Muslims to join the terrorist sectors of the religion, abuses he himself encountered:

> Our sexual desires were dramatically stimulated from reading Islamic books. The following Hadith and Quranic verses illustrate how these books increased our sexual curiosity:

(78:33) Young women with pubertal breasts are waiting for you in paradise.

(55:72) Ladies with beautiful, big, and lustrous eyes are waiting for you inside the tents (in paradise).

The over-stimulated sexual desires of young Muslims...the hopelessness in soon having a marital relationship, and dreams of beautiful women waiting in paradise engender frustration, anxiety and anger. These factors encourage young Muslim men to join radical Islamic groups where they then become steeped in terrorist Islamic beliefs such as committing suicidal attacks on infidels to go immediately to paradise as martyrs so they can enjoy the beautiful ladies there, especially the 72 virgins.[37]

Taking the extreme requirements of Islamic submission to their more hideous limits, it is also necessary to illuminate the religion's practice of castration. Girls and women are castrated in many Muslim societies by female genital mutilation (FGM). UNICEF estimated that in 2016 upwards of two-hundred million women in thirty countries from Africa to Asia to the Middle East were subjected to this horrifying, debasing manner of castration. Not only does FGM act to diminish women's sexual drive, it subjugates them to their cultural superiors in the most intimate of ways. Ironically—and perplexing to the Western mind—the practice is perpetrated often by older women and is considered a ritual of purity and modesty and is requisite for social acceptance. Therefore, to not comply, a girl would risk being shunned, or worse. The pressures of the culture align to keep her in her place, as the lowest figure in Muslim social ranking.[38]

Men, too, have historically been castrated in Muslim societies— though especially foreign men, those conquested in wars of aggression or those taken as slaves. During the Ottoman expansion of the twelfth century into southeast Europe, Christian boys and young men faced the very real threat not only of conscription, but of being made into eunuchs and pressed into service to the elite military ranks. Eunuchs, of course, posed no threats to the men they served, presented no competition for the women, and were considered disposable if they did not do their jobs.[39]

The peak-testosterone philosophy of the Hebrew Bible, by distinction, is highly potent and forbids any form of castration as a pagan practice that leads to impotency, weakness, submission, and slavery. Castration contradicts the very purpose of creating life. "Judaism has always forbidden all forms of castration. Alone among the nations of antiquity, the Hebrews imposed a religious prohibition on the emasculation of men and even animals, a prohibition not found in the teachings of Buddha, Confucius, Christ, or Muhammad."[40]

## How Group Conflict Leads to Fascism and Communism

The decline in testosterone levels triggers individuals to coalesce into groups in order to adjust to the changing environmental conditions of low energy and constrained food supplies. The transformation in how society is organized leads humans to make the required adjustment to more adversarial living conditions by reducing reproductive activity and limiting birth rates through population control. Society moves to a tightly ordered, regimented formation and becomes susceptible to a herd mentality and a despotic leader. People are readying for group aggression and wars for resources. Group conflict ensues as humans lose their individualism to become beasts of prey, social animals that form rival coalitions that battle over reduced available resources and supplies. The political manifestation of this is the rise of authoritarian regimes such as communism and fascism and the concurrent decline of liberty, individual rights, and free-market capitalism.

Adolf Hitler's lack of testosterone can be confirmed by the knowledge of his physiology in his condition of micropenis. Low testosterone *in utero* is the cause of this lack of development. Furthermore, he had only one testicle due to a medical condition called hypospadias, and was therefore impotent. [41] One of Hitler's compatriots, Ernst Hanfstaengl, described his impressions of Hitler's sexuality in writing, "I felt Hitler was a case of a man who was neither fish, flesh nor fowl, neither fully homosexual nor fully heterosexual ..."[42] His personal condition serves as further evidence for the Testosterone Hypothesis' claim that testosterone collapse during the 1920s and 1930s led to the rise of the death cult of National Socialism in Germany.

The German creed of Nazi Socialism self-identified as a pagan masculine cult, referring to the German Aryan race as the master race and their territory as the fatherland. The Soviet communist regime, by contrast, called Russia the motherland. This develops into sadomasochistic relations of dominance and submission through bondage. Even in a lower testosterone environment, such as in a lesbian relationship, in which the "butch" partner with the more masculine characteristics tends to dominate the "femme" partner with feminine characteristics, testosterone drives gender role differences. [43] At the global level, the testosterone variances revealed themselves in the aggressiveness of the Germans against the Russians. The Nazis believed their superorganism required the conquest of more territory in order to provide for *lebensraum*, or living space, that would provide the resources to expand their super race by enslaving the inferior races. The Soviet communists, by contrast, had an extreme feminine, egalitarian vision of submitting humanity to a collectivist utopia in which all are as one. The Nazis sought to expand their race, or genetic makeup, at the expense of the other races, because they viewed their culture as superior; while the communists sought to mix the gene pool of humanity under the egalitarian social vision. So it was that the more dominant Nazis attacked the Soviet communists. Both were antagonistic to the Judeo-Christian, biblical, patriarchal vision of individualism, family values, and personal liberty.

When testosterone levels change, the dynamics of the dominance hierarchies lead to social conflict over the new in-group rankings. Societies are groups of individual members among whom there are intragroup conflicts; but there are also intergroup conflicts on the level of greater social groups, such as nations and religions, which are part of the entire human species. As we discussed in the origins of sex, cultural evolution is driven by the biological imperative to mix genes in order spread the genetic variations in the species. Furthermore, the conception has been popularized that our culture evolves through the exchange of ideas, similar to the mechanism of biological evolution through the exchange of genes:

> Sex evolved because the benefit of the diversity created through the intermixture of genomes outweighed the costs of engaging in it, and so we enjoy exchanging our genes with one another, and life is all the richer for it. Likewise ideas. "Exchange is to cultural evolution as sex is to biological evolution," Ridley writes, and "the more human beings diversified as consumers and specialized as producers, and the more they then exchanged, the better off they have been, are and will be. And the good news is that there is no inevitable end to this process. The more people are drawn into the global division of labour, the more people can specialize and exchange, the wealthier we will all be."[44]

There are periods of equilibrium and peace in which a dominance hierarchy is stable and the group maintains its integrity and social order. However, to allow for greater intermixing of gene sequences to spread among the population of the species, punctuated periods of disequilibrium, meaning chaos and war, are required to change the dominance hierarchies. The biological imperative to achieve the dynamism that reshuffles the gene pool of the species is the root of recurring cycles of war and conflict as one order and social hierarchy is eliminated to be swapped with a new.

### The Significance of Height

Having noted the peculiar physiology of Adolf Hitler, it is interesting to recognize that other infamous dictators also present with physical characteristics that indicate low testosterone. French Emperor Napoleon Bonaparte also had a micropenis; and of course, Napoleon's stature and temperament combined became the archetype for the complex bearing his name. Hyperbolic machismo and narcissistic power plays, along with blinding jealousy that is bred by insecurity, all characterize the "Napoleon Complex" that some short men are apparently driven by to compensate for their size. In addition to Napoleon himself, Stalin, Mussolini, Attila the Hun, and Nero were infamously rabid in their leadership—and were notably short men.

In so far as perception can be measured, studies indicate that people prefer to choose tall people as leaders when given the choice. Author Malcolm Gladwell reports that fifty-eight percent of the top executives of successful US companies were at least six feet tall, whereas only fourteen percent of the population is that tall. He goes on to say that since the era of movies and television, when people could actually see the politicians running for office, the taller candidate has won three-quarters of the US presidential races over a shorter candidate. Yet even before people necessarily saw their candidates in moving pictures, the pro-liberty presidents of the United States were also the tallest, suggesting high-growth hormone levels: Abraham Lincoln, Thomas Jefferson, George Washington. Of the forty-five presidents who have served the US, fifty-six percent of them were five feet eleven or taller.[45]

While Americans might vote for taller candidates, unfortunately, Americans themselves are all but shrinking. This is to say that over the past fifty years, the average heights of both men and women have stagnated in the US. Data compiled by the NCD Risk Factor Collaboration shows that the population born after 1960 is not increasing in height, a fact that reflects declines in growth hormones in the populace. In the twentieth century, America went from number three in the world for height to number forty. According to the researchers, genetics do not account for this change nearly as much as diet, health, and environment.[46] A culture that has been struggling within itself during that same time period for its very ideological identity reflects that struggle in the hormonal profile that determines height, among other things.

Furthermore, height is a significant metric for societal prosperity. Writer Dan Kopf states, "Countries with tall people are wealthier, have longer average life spans, and are less likely to have experienced conflict. There's no better sign of a country's health and wealth than height."[47] He goes on to cite:

> Other anthropometric researchers go so far as to suggest that height and height growth are a better measure of human development than Gross National Product (GNP). "GNP doesn't have anything to say about inequality; height does," economic historian John Komlos told *Scientific American*. "GNP has nothing to say about children; height does. So it reaches the part of the population that's left out of GNP measures."[48]

These economics reveal themselves strikingly in the country with the greatest height rate growth of the past decades: South Korea. A nation that has embraced free market economics and prospered significantly since the 1960s, following their civil war, has also experienced the greatest growth rate. Moreover, South Koreans have grown to be an average of 1.2 inches taller than their genetic brethren in North Korea in the same time period.[49]

## How Testosterone Regulates Breeding

If we study the history of our civilization, we see recurring, cyclical episodes of cultures growing to be become great nations, only to fall down again under barbarian conquest, rape, and submission. To recall the evolution of our Judeo-Christian heritage, we look into its past to find that ancient Israel became a great nation after the Exodus from Egypt led by Moses, peaking in dominance and independence with the ideal of liberty and lack of central government during the period of the judges for two centuries. This peak testosterone, heterosexual, patriarchal culture, which followed the Father God figure, rejected foreign cultures as pagan idolaters and maintained a strong internal identity. Ancient Israel defined a closed-borders policy, territorial integrity, and a very low rate of interbreeding with foreign social groups. However, as testosterone levels declined, the Israelites succumbed to a king and centralized authority like the pagan nations surrounding them. During the successive generations, the Bible tells us that King Solomon intermarried with hundreds of foreign wives and concubines and built pagan temples, mixing foreign genes along with ideologies and religions in with the Israelites. Hence, testosterone regulates social group identification of in-group versus out-group behavior. When testosterone is high the males and the females have an in-group preference, and social homogeneity and cohesion is high, but such preference declines with lower levels.

By 930 BCE, decline in testosterone levels led to group conflict over the dominance hierarchy in the Hebrew kingdom. Cultural homogeneity and group cohesion could no longer be maintained, as the nation split into two separate kingdoms, Israel and Judea. Both of them would be conquered and raped by the Assyrians and the Babylonians over the next few centuries, with their gene pools and ideologies intermingled with the pagan cultures in the territory around them.

Christianity was created by the intermixing of Judaism and pagan Greek philosophy, first under Hellenistic Judaism after the empire created by Alexander the Great, and then under the unifying force of the Roman Empire that conquered the entire territory. The pagan cultures, in contrast to the peak-testosterone biblical creed, more willingly embraced each other's idols and swapped genes and ideas in the aim of unifying humanity under a single, encompassing empire.

Therefore, when testosterone declines, we can observe in history that territorial boundaries and independent nation states fall under the conquest of global imperial forces, such as Alexander the Great, the Roman Empire, Islam, the Mongol Hordes, Napoleon, or the Soviet empire and National Socialism in the twentieth century. The great civilizations in their turn go through these hormonal cycles of rising to dominance and independence under a republican form of government with law and order, to gradually degenerate as they become less masculine into a centralized imperial form. Finally, they fall down in submission to conquest by foreign invaders.

To observe such a plummet into submission can be perplexing. In the case of the Jewish people in the past hundred years, why is it that so many have willingly become socialists and communists in compliance to overbearing governmental authority? Given their rich historical legacy of the peak-testosterone, dominance philosophy of the Hebrew Bible that commands liberty and independence from pagan oppression and bondage to the Pharaoh in Egypt, as well as the more recent history of such horrors under fascism, this inclination seems incredible.

Friedrich Nietzsche, in his ideas of the animalistic nature of man and his fascist ideal of the superman, understood the fall from dominance in the collective psyche of the Jews during their subjugation under the Roman Empire. The ensuing "slave morality" caused them to invert their moral systems and eventually incline them toward cultures of submission like communism. Vilifying the masters, such as the Roman occupiers of ancient Palestine, by slaves devalues strength and wealth and succumbs the oppressed people to an ethic of weakness. In Nietzsche's schema, such weakness makes poverty and humility a virtue

> By saying humility is voluntary, slave morality avoids admitting that their humility was in the beginning forced upon them by a master. Biblical principles of turning the other cheek, humility, charity, and pity are the result of universalizing the plight of the slave onto all humankind, and thus enslaving the masters as well.[50]

The Jews who go along with this inverted morality are those prone to communism and other cultures where they are subject to domineering, "master" authorities.

### The Alpha Male, Monogamy, and Infanticide

Yet another aspect of the Western males' fall from dominance is the injurious effect on their children. Successive generations are compromised when foreign invaders gradually conquer territory and enslave the population. In many mammals, such as lions, when the alpha male ages and loses his potency and ability to defend his females in the pride from younger and stronger male competitors, the results are often disastrous for his young offspring. The new male will battle and expel the old one, and will often kill the young cubs of the former king. As one researcher puts it, the lions "can't be stepfathers." Therefore, to kill the existing young frees up the females to mate and allows the dominant male to pass on his DNA.[51]

In fact, it has been postulated recently that the evolution of monogamy in primates and possibly humans was in part a way to safeguard the species' progeny. Primates exhibit pair-bonded monogamy at a rate of twenty-seven percent compared to a mere five percent in other mammals taken together. Oxford University scientists claim that, given their larger brains that take longer to develop, ape young are more vulnerable for longer. Their need for protection led to the relative relational stability of the parents. This was a "major evolutionary transition, which dramatically altered the evolutionary trajectory of our species," according to the researchers.[52]

Furthermore, there is evidence that humans are also genetically disinclined to being stepparents, making children more susceptible to abuse when the biological parents are not the primary caretakers. When the dominant biological father is absent from taking care of his children evidence suggests that their well-being is compromised:

> Family structure is the most important risk factor in child abuse and infanticide. Children who live with both their natural (biological) parents are at low risk for abuse. The risk increases when children live with step-parents or with a single parent. Children living without either parent (foster children) are 10 times more likely to be abused than children that live with both biological parents.
>
> Children who live with a single parent that has a live-in partner are at the highest risk: they are 20 times more likely to be victims of child abuse than children living with both biological parents.[53]

# 2 • DOMINANCE AND SUBMISSION

In addition, males are less secure in their paternity and paternal investment, particularly in a promiscuous culture in which the females might be fertilized by men other than their husbands. Among other factors, the feminist culture of divorce drives men out of their children's lives, ending the fathers' natural masculine role as providers and protectors. This causes the children to become exceeding vulnerable to sexual exploitation from other men, including pedophiles. Another symptom of this low-testosterone cultural change is a movement to lower the age of sexual consent, which allows for pedophilia. There is an infantile quality to a society that cannot make the distinction between a mature person and a child. In France, for instance, more than once the courts have declined to charge an adult male with the rape of eleven-year-old girls "because authorities couldn't prove coercion."[54]

Hence, family values are only promoted in the patriarchal tradition of the Bible, as the Ten Commandments prescribe against adulatory. In periods of declining male dominance, such as the feminist sexual revolution of the 1960s, the iconic women's liberation movement song "You Don't Own Me," presented a rebellion against male authority with the lines, *You don't own me/I'm not just one of your many toys./You don't own me/Don't say I can't go with other boys.*

As described before, Islamic nations, such as Iran, which were previously under secular regimes, consequently went through an Islamic revolution during the late 1970s, leading to a counter-reaction of sexual repression, depression, and submission of women with the mandatory wearing of the hijab. This is probably a result of collapsing testosterone levels, resulting in a loss of male self-confidence and social status and inclining them to organize for suppression and control of female sex drive.

### Genetic Battle of the Sexes

Pair-bonding and commitment to monogamy has the advantage of aligning the interests of the male and female for raising the offspring. However, in the increasingly promiscuous, sexually liberated feminist culture producing high divorce rates, there are growing conflicts of interests between parents regarding their mutual children. A new theory of the genetic battle of the sexes might explain how this is causing more and more children to be diagnosed with autism and schizophrenia.[55][56]

The imprinted brain theory originates in evolutionary psychology, determining the causes of autism spectrum disorders and psychosis. Genomic imprinting is an epigenetic process in which certain genes, inherited from either parent, are expressed in the offspring in a different manner according to their origin. The imprinted brain theory, a variant of the conflict theory of imprinting, suggests that in humans the maternal versus paternal sets of genes might exhibit antagonistic reproductive interests as they battle for dominant expression in the offspring. In the 2009 book *The Imprinted Brain: How Genes Set the Balance between Autism and Psychosis*, Christopher Badcock argues that symptoms of psychosis can be shown to be the mental mirror-images of those of autism.[57] The extreme male brain tends toward autism, while the extreme female brain tends toward psychosis:

> The female brain tends toward empathizing and mentalizing thinking, treating machines and objects as if they were other people. They attribute minds, thoughts, and feelings to inanimate objects. That, according to Crespi and Badcock, is the essence of paranoid schizophrenia. Paranoid schizophrenics hear voices where there are no people, and they attribute minds and thinking where none exist, such as when they believe other people are talking about or conspiring against them when they aren't. Paranoid schizophrenics are hypermentalistic, and overinfer minds and emotions in other people, just as autistics are hypomentalistic, and underinfer minds and emotions in other people.[58]

It is intriguing to consider that such a hyper-feminized brain of paranoid schizophrenic behavior might explain the extreme feminist aggressive stance against Western male dominance. Insomuch as they view all males conspiring to subjugate women, they deny it was Western ideals of liberty that liberated both men and women and provided them with equal rights under the law. This is similar to Hitler's Nazi cult of anti-Semitism that blamed the Jewish patriarchy with a global conspiracy to exploit the German people and wage war against them, when it was actually the Judeo-Christian ideal of liberty and independence that led to the European Enlightenment culture of individual rights. The Enlightenment English Puritan philosopher John Locke, the father of classical liberalism in the modern age, was a Unitarian who identified with the ideal of God the Father of liberty. He derived his ideas of fundamental human equality and liberty for both genders on the verses from Genesis 1:26-28, the beginning of the doctrine that both male and female were created in the image of God.[59]

### Phase Transition in Cultural Evolution

Testosterone collapse brings the death cycle of civilization in phases. This phenomenon is thoroughly examined in *The Testosterone Hypothesis*, but here we will look briefly at the second half of the twentieth century, following World War II, and identify three phases of transition:

1. 1950s: High testosterone, masculine leadership creates a culture of family values, high birth rates, capitalist prosperity, and peace.

2. 1970s – 2000: Testosterone decline brings the feminist revolution that destroys families, attacks individualism, and increases dependence on big government and socialism. Testosterone collapse in the 1960s led to the feminist–socialist revolution against the male patriarchy and the increase in non-western immigration into the West.

Hormone cycles regulate rates of ethnic diversity, interbreeding, and multiculturalism. In Sweden in 1975, the parliament liberalized immigration policy, which led to growing waves of third-world migration and a gradual change of the former homogeneous culture into a multicultural society. A similar path was taken in the United States in the 1960s and thereafter. Canadian and European immigration dropped almost forty percent between the 1950s and

1970s, while Asian immigrants came to North America at a rate that increased twenty percent in the same time frame.[60]

3. 2000s: Testosterone collapse brings the feminized culture to finally fall down in submission to Islamic Jihad and ultimately may well lead to a nuclear Armageddon in WWIII.

In the new millennium the rapid fall of testosterone levels has brought about a host of symptoms that feed the vicious cycle of low social mood and societal collapse: Depression is up, along with more men leading sedentary lifestyles, becoming addicted to drugs and alcohol at alarming rates, and foregoing procreative relationships with women. This can mean, as noted in chapter 1, that fewer men are sexually active or are finding their sexual activity through pornography. But it also manifests in a parallel trend of greatly increased promiscuity among people in their child-bearing years whose casual sexual encounters are mitigated by birth control and abortion, all leading to lower fertility rates across Western societies. Men might be having sex, but their decreased testosterone deters them from competing for dominance in actual *mating*, in actually producing offspring and furthering their DNA.[61]

Furthermore, men are physically weaker and less equipped to protect their families. Even though the stabilizing benefits of Western liberalism and individual rights have given the world more wealth and prosperity and law and order that has reduced violence, threats, and aggression in many places, the disadvantage to this condition is that Western men have lost their aggressive instincts to protect their territory, nations, and women from foreign enemies. The balance between testosterone and cortisol is a key factor in motivating men to pursue and maintain status in social hierarchies, so when that is askew, only the truly domineering males maintain power.[62]

Such a cascade of negative symptoms naturally leads some to consider that masculinity itself is "dying" in our societies. In a headline, writer John Stonestreet asks, "Where Have All the Men Gone?" and proceeds to describe the state of things in our culture:

> Most men today aren't just physically weaker than previous generations. They're weaker as people: guys

stricken with "Peter Pan syndrome" never leaving adolescence; "safe spaces" on college campuses protecting perpetually fragile constant-victims from serious debate; the hookup culture and porn addiction replacing chivalry; and the sexual revolution promulgating through media, education, and now law that there's no such thing as male and female. What we're seeing isn't a different expression of masculinity adapting to new cultural realities. What we're seeing is no masculinity at all. [63]

What Stonestreet is lamenting can be further evidenced in more outlying exhibitions of the feminized Western male becoming submissive—submissive to the point of encouraging their own wives to be adulterous. The cuckold is a husband who masochistically drives his wife to sleep with other men. The man actually finds pleasure in being humiliated. Psychologist Roy Baumeister suggests in *Masochism and the Self* that cuckolding, among other manifestations of sexual masochism, is an avenue for a man to relieve himself of the stressful burdens of reproduction and retreat into a less expansive role. The number of Internet searches for this topic have more than doubled in the past decade, and online groups supporting the practice are thriving. [64]

Cultural evolution is operating through all of these factors. As the man in the civilized West is becoming impotent, it is the barbarian hordes that will conquer them and rape and fertilize their women. This has been the course in all the cycles of civilization's history. One of the great conquerors and empire builders was the Mongol chief, Genghis Khan, who ferociously gained domination over the landmass stretching across most of Eurasia. He describes "the greatest happiness" in these terms, "[T]o scatter your enemy and drive him before you. To see his cities reduced to ashes. To see those who love him shrouded and in tears. And to gather to your bosom his wives and daughters." [65]

The barbarian hordes gain the upper hand when civilization is weak from within due to testosterone collapse that eradicates masculinity. Currently in the West, the feminists are in the final phase of castrating, eliminating, and replacing the impotent Western male with the barbarian hordes of Islamic jihad. The feminists seek to interbreed with the Islamists, under the multicultural agenda, because they are viewed as liberators from Western male dominance. In the wake of one dominance hierarchy falling, a new one will emerge to take its place. The swapping of genes across race and ethnic group lines will be achieved to spread the genetic variation through the population of the human species. Such, it seems, is the way of the world.

# 3

⋯⋯◗◖⋯⋯

# CULTURAL EVOLUTION AND THE GENERATION GAP

*All mankind together make continual progress in proportion as the world grows older, since the same thing happens in the succession of men as in the different ages of single individuals. So that the whole succession of men, during the course of many ages, should be considered as a single man who subsists forever and learns continually.*[1]
— Blaise Pascal

*Society is indeed a contract. It is a partnership . . . not only between those who are living, but between those who are living, those who are dead, and those who are to be born.*[2]
— Edmund Burke

*Each new generation born is in effect an invasion of civilization by little barbarians, who must be civilized before it is too late.*[3]
— Thomas Sowell

The evolution of human civilization is an historical process that involves the rise and fall of human societies, initially rising to develop independent cultures, traditions, philosophies, religions, and social systems, but later—usually through conquest and aggression—merging with other social groups, integrating culturally, and creating a new social system.[iv] We see that history gives birth to great civilizations that reach maturity, gradually age and die out, and are subsequently replaced by something new. Unless the old cultures face total extinction, both the biological information in their genes and the knowledge they acquired are transferred though cultural assimilation into the body of the emerging society that follows them.

Human cultures are similar to human bodies in their dependence on information for their social organization and ability to adapt. Biological evolution works according to the programs set by the genes, in the multitude of cells in the human body, designed to adapt in response to changing environmental dynamics. Cultural evolution is the adaptation of human groups, composed of many individual members, to changing environmental conditions. But why is it that individuals and human civilizations age and die? A mature living organism cannot be fundamentally transformed, as its structure is already built in and cannot be altered in a meaningful way. Hence, evolution introduced the process of aging and death to allow for the reproduction and growth of a new generation that will be better adapted to the new environment.

## The Generation Gap

The familiar term *generation gap* was coined in the 1960s to refer to differences in opinions, values, and practices between younger people and older people. The sociological theory was developed as the post-war Baby Boom generation grew to rebel against the values represented by its parents. Including the anti-war movement and the sexual revolution, this turbulent decade found both family values and capitalism attacked by the women's liberation movement and the growing trend toward socialism. The numerous and often violent protests that characterized this particular generation gap offer familiar historical images to us even fifty years later. The reality is, however,

---

[iv] For a detailed treatment of the historical cycles or "waves" of Western civilization, please refer to my previous book *The Objective Bible* (2014). See especially "Part 3: History of Civilization as Evolution."

that gaps can be perceived between most generations—although some are more pronounced than others.

Sociologists divide the human lifespan into three distinct phases: childhood, maturity, and retirement. Each age group is designed to socialize with its own members and has less interaction with or influence from other age groups, except at the level of the nuclear family. In the West, the generations of the past century are denoted by birth years and nick names:

1901-1920: The "Greatest Generation" came to age during the Great Depression, living through and fighting in WWII.

1925-1941: The "Silent Generation" focused on careers and traditional social norms.

1946-1964: The "Baby Boomer Generation" was born in a period of increasing fertility rates and is famous for its nonconformity during the 1960s and 70s.

1965-1980: "Generation X" matured into adolescence during the 1980s and into the 90s, having entrepreneurial tendencies.

1981-2000: The "Millennial Generation" matured during the depression of 2008 and its after effects.

2001-Today: "Generation Y" is growing through a period of testosterone decline leading to the great recession and emotional insecurity.

## Puberty and Adolescence

*Puberty* is the growth phase of a child in which a process of physical changes brings the maturity of the body into the adult stage of development, finally acquiring the capabilities of sexual reproduction. The initiation of puberty is regulated by the rise of growth and sex hormonal signals from the brain to the gonads: testicles in boys, and ovaries in girls. The gonads, in turn, respond with increased hormonal production, stimulating the growth and transformation of organs in the body, such as the brain, muscles, bones, breasts, and sex organs, as well as the acceleration of physical growth of body mass and height.

Significant structural or morphological changes are the secondary sex characteristics that appear in sexual maturity, particularly the traits that distinguish the behavior of the genders, but are not strictly a part of the reproductive system itself. These characteristics are a product of sexual selection for traits that increase reproductive fitness in social animals, supporting the individual success in acquiring mates through strategies such as aggressive behavior, dominance, and courtship.

*Adolescence* is the period of cognitive skill development in the transition from childhood to adulthood that overlaps puberty. Because of rising testosterone levels in this phase, teenagers are known to be rebellious against authority and to seek independence from their parents in developing their own character. Additionally, the hormonal changes incline both male and female teens to high-risk social behavior.

Life cycles are determined by our endocrine system as affected by low levels of solar activity in the sun's radiation cycles. While this will be more thoroughly discussed in chapters 7 and 8, generally speaking, as the sun releases less energy, our body responds by releasing fewer growth hormones. When testosterone levels are depressed society wide, it leads to an infantile generation that fails during puberty to gain the maturity, independence, and self-determination to take control of life. Consequently it seeks to remain servile under a paternalistic system. Paternalism is the behavior of a centralized organization, such as the state, that limits people's autonomy and liberty by determining and enforcing the proper standards for a good society from above. This implies that a person lacks the free will and rationality to take responsibility for his or her own actions, requiring adult supervision and control by authority to regulate behavior.

If testosterone levels remain subdued for a long period of time, the next generation going through puberty will continue to grow into adolescence with an acute deficiency in individualistic personality, and the culture will degenerate into a state of depression. The older generation, having developed an independent character during a period of higher testosterone levels, will gradually be replaced with the new generation that is mentally and culturally wired to choose a different system of social organization.

Today's younger generation is greatly affected by the current decline in growth hormones. Whether that is evidenced in their politics or their grooming habits, the differences are evident. Millennial males have adopted styling preferences that were once relegated to the homosexual community. The term "metrosexual" was coined in the 1990s to describe predominantly heterosexual, urban men who give a lot of attention to their appearances. This has melded into even more widespread "hipster" styles that focuses on pristine grooming of men's facial hair, including eyebrows, and distinct fashion preferences. The more feminized heterosexual men have taken style and grooming to new levels, while at the same time tending to be more docile, passive, and submissive to authority given their lack of self-confidence and an internal locus of control.

Testosterone collapse causes social anxiety, depression, and risk-aversion, which we now see manifested in the rise of populist socialist movements that promise a kind of caretaking and are led by controlling, authoritative leaders. It is no wonder that the millennial generation became so excited about Bernie Sander's communist, big government, paternalist agenda in the 2016 presidential election cycle. Only thirty-eight percent of millenials and Gen Y young people consider communism "very unfavorable;" whereas fifty-seven percent of the culture at large finds it so. When queried about socialism, the split looks like fifteen percent and thirty percent unfavorable respectively.[4] The Victims of Communism Memorial Foundation reports:

> The survey found a similar dynamic around capitalism. Only 42 per cent of millennials have a positive view of the system, compared to 54 per cent of baby boomers and 71 per cent of mature adults.
> The survey found that younger respondents were less receptive of ideas proffered by minds such as free market economist Milton Friedman, who said, "A society that puts equality before freedom will get neither. A society

that puts freedom before equality will get a high degree of both."

Younger people were more receptive to Bernie Sanders's ideas: "A nation will not survive morally or economically when so few have so much, while so many have so little."[5]

Another study shows that millenials are more ideologically aligned with the teachings of Marx than those of the Bible.[6] It makes sense that high stress hormone (cortisol) levels that lead to dropped testosterone levels compel people to prefer safety and conformity under a communist central planning authority rather than taking entrepreneurial risks and pursuing innovation in the capitalist economic system.[7] Even millenial generation conservatives report more liberal attitudes regarding sexuality and socialism, as depicted in the chart in Figure 5.

**Millennial Republicans More Liberal than Older Republicans on Homosexuality, Immigration**

*% of Republicans and Republican leaning independents in each generation who say...*

All Rep/RL　　Millennial　　Gen X　　Boomer　　Silent

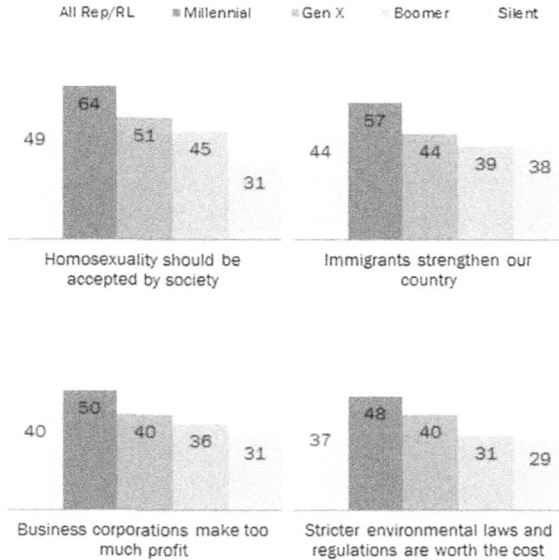

Homosexuality should be accepted by society
49 · 64 · 51 · 45 · 31

Immigrants strengthen our country
44 · 57 · 44 · 39 · 38

Business corporations make too much profit
40 · 50 · 40 · 36 · 31

Stricter environmental laws and regulations are worth the cost
37 · 48 · 40 · 31 · 29

Figure 5: Millennial Republicans more Liberal [8]

The political tendencies of today's younger generation are infamously accompanied by pervasive hypersensitivity, gaining them the labels "Snowflakes" or "Crybullies" as they throw public tantrums in their inability to engage with conflicting opinions and ideologies.[9] Acutely self-centered, they insist on any and every version of gender identity, their very own pronouns, safe spaces and coddling in their schools and universities, and universal crack-downs on any speech considered offensive.[10] When millenials band together *en masse*, they create a harrowing mob that throws manipulative—even violent—fits that invoke authorities to capitulate to all manner of demands. They epitomize the irony that stress and fear exacerbated by high cortisol overtake the creative competition and independent courage of healthy testosterone levels. A bit of anecdotal evidence for the low testosterone levels of millenials comes from the popular, liberal-leaning news website BuzzFeed. Four of the men who work for the site had their testosterone tested, and all of their levels were well below the average for their age group. In fact, their testosterone was below what an average 85-year-old's levels would be![11] While the sample size is very small, the results nevertheless align with the trend that has already been cited: testosterone levels are declining precipitously, and particularly in the younger generation that inclines toward the ideological left.

In our time we see civilization rapidly changing before our eyes. The older generation, with its high-testosterone, masculine culture of independence, strong family values, and higher birth rates, is aging. It is being replaced with a young generation that rejects Western capitalism and the older generation's commitment to traditional family values, resulting in a declining phase of civilization. We see the millennial generation losing its drive for life and independence, losing its gender identity and will to reproduce, and being more passive and submissive to increasing governmental authority, seeking to be increasingly controlled and governed by a Big Brother type socialist regime.

The social characteristics of the old generation were depicted well by the German sociologist Max Weber in his 1905 book, *The Protestant Ethic and the Spirit of Capitalism.* Weber describes the Protestant (particularly Calvinist) work ethic, strong family values, and the economics of high savings rates and production as the major forces behind the rise of capitalism. The Calvinists based their worldview on biblical values and a patriarchal society.

As testosterone declined in the late nineteenth century, Karl Marx's communist collaborator, Friedrich Engels, attacked the institute of the family in his 1884 book, *The Origin of the Family,*

*Private Property, and the State* as the organ of inequality and enslavement. The communists understood that man's natural desire to care for his own progeny, his children (to whom he passes on his genes), through the generational accumulation and inheritance of private property, is at the root of the desire for the private property rights and the capitalist system. Engels related this drive to the rise of monogamy and the institution of marriage in which a man secures his reproductive rights with his woman as "an instrument for the production of children."[12] Hence, Engel refers to marriage as the enslavement of women and origin of inequality that will only end in a communist utopia, in which the state collectively owns all the means of production.[13] By terminating man's rights to private property, communism is meant to lead men to cease selfishly caring for his own offspring. Further, it would lead to the end of the institution of marriage and open the gates for free sex.

Testosterone continued its decline in the 1920s, during which time the Frankfurt School developed in Germany to build the movement of cultural Marxism. This philosophical paradigm led to further destruction of the family in order to usher in total collectivism and communism.

## How Sex Determination Shapes the Epochs of History

In this chapter we are looking at how changes from generation to generation impact the cultural evolution of societies. In chapter 2, the biological phenomena of male and female characteristics were discussed. Females are designed to exhibit their fertility and symmetry in order to attract male suitors, therefore, their brains are more inclined for interest in esthetics, beauty, and design. The male brain is more interested in acquiring and creating resources to elevate its social status in order to compete for females. The androgynous personality type, however, has also been very influential throughout the course of human history. Art historian Camille Paglia has discovered recurring patterns in which "Sexual Personae" shape culture, art, and world philosophy, manifesting sexual characteristics into every facet of human social life. For instance, the homosexual cultures of ancient Greece and Renaissance Florence produced great art, because the homosexual brain is an androgynous type that is organized in utero as a female brain. Exposure to low testosterone levels in utero drives this brain type to be attracted to males and have greater interest in art and beauty; but during puberty and adulthood, homosexuals have the same testosterone levels as all males, and hence have a great competitive drive for creativity. Such a mixed brain type is responsible for the great artistic achievements of these historical epochs, such as the *Mona Lisa*, an androgynous depiction of immense beauty. In contrast to the pagan, feminine brain-type cultures that worship the body, the peak-testosterone culture of ancient Israel was concerned with creation of the material world through language, as God created the universe by speech in Genesis 1, and he condemned the worship of nature or material objects as idolatry.

The Organizational-Activational Hypothesis states that there are two significant periods in which the steroid hormones are responsible for the sexual characteristics in animal organisms. In early development sex hormones impel the permanent organization of the nervous system, and upon reaching adulthood these hormones again activate and modulate these behavioral traits. The biological system that governs the sexual characteristics developed in an organism is called the sex-determination system. Because we are all created female in the default state, it is the rise of testosterone in utero and later in puberty that determines the degree of sexual differentiation in our society. During high-testosterone periods in which sexual differentiation is maximal, we think of our sexuality as binary—having two distinct sexes of male and female—determined by the existence of two X chromosomes in females, versus one X chromosome and one Y chromosome in males. However, studying the evolution of sex suggests that sex chromosomes developed in a later period of evolution, and initially there were other mechanisms for sex determination. Some species do not have a fixed sex, and instead change sex based on certain cues. There are many other sex-determination systems such as temperature dependent sexuality in some species of reptiles, such as alligators and some turtles, in which sex is determined according to incubation temperature of the egg during the sensitive period.

The various ways in which sex hormones shape our cultures today are a product of this complex system of sex-determination that evolved over hundreds of millions of years. In order to understand the forces of history that shape our cultural evolution we must study biological evolution as major driver of our social transformation.

## Stress

As Millenials and many of their Baby Boomer parents actively try to usher in the nanny state, they seem to be ignorant of the strong Western culture they are working to supplant. A difficult fact of our culture's success and abundance is the hormone-driven tendency in mankind to cycle out of our own prosperity. Hormonally we are actually propelled by degrees of stress, and when the "good" kinds of stressors give way to the myriad depleting stressors, we degenerate.

### Eustress versus distress

A beneficial or healthy stress level is called *eustress* (*eu* meaning good). This term pertains to either psychological or physical efforts related to desirable events and positive feelings of fulfillment in the course of a person's life. *Distress*, in contrast, is a type of stress that has negative implications.

The Yerkes–Dodson law was developed in 1908 denoting the relationship physical between arousal and performance, demonstrated empirically. This law refers to increased performance according to levels of mental and physiological arousal. After a certain limit, however, conditions of over-stress lead to exhaustion and a decrease in performance. Figure 6 shows a graphical illustration of this process in the form of a bell curve, where the relationship can be seen between effort and arousal, or motivating inclination. Once the tipping point is reached, compelling arousal becomes negative anxiety, increasing as performance wanes.

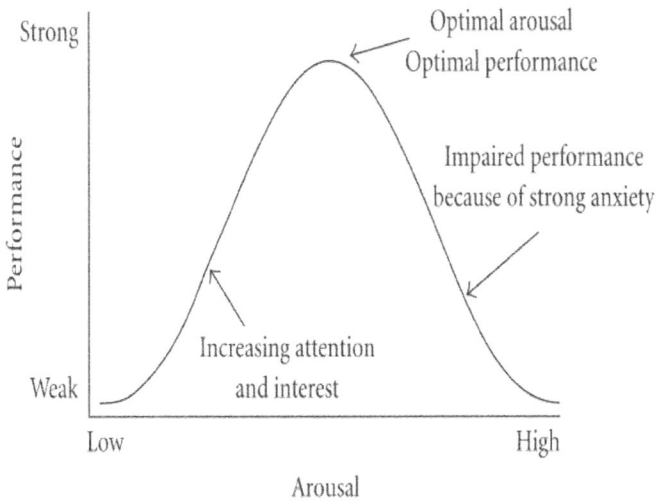

Figure 6: Correlation between Arousal and Performance [14]

The beneficial effects of stress can come in various forms. It has long been known that certain "poisons" administered at just the right dosage can induce the body to produce healing agents. Exceeding that dosage instigates harmful effects. This is termed *hormesis*. Stress-response hormesis refers to the phenomenon in realms other than toxicity. Low levels of allergens such as cat dander are given people with cat allergies to stimulate their immune response. Likewise, pesticides used in small amounts can engender pest resistance in plants. The principle of hormesis is now being applied in studies on aging in various organisms.[15]

A 2007 study of the effects stress hormones (glucocorticoids) have on cognition discovered that the relationship expressed in the Yerkes–Dodson curve is also manifested in tests of memory performance. To fully optimize the potential in forming long-term memories, one should mildly elevate glucocorticoid levels, whereas overabundance of stress hormones reduces the formation of long term memories.[16]

One example of positive stress is physical effort, such as exercise or resistance training, which causes stress on the muscles that lead the body to build strength and power to do work. In the same manner that evolution is an adaptive process, our body is designed to adapt itself to our environment and living conditions. Unless they are propelled to do work, our muscles will atrophy, declining in size and strength and becoming weak to the point of disease and even death. Exercise and physical work raises both our stress hormone, cortisol, to break down fat and supply the muscles with sugar, and our anabolic hormone, testosterone, to grow the muscles. This is a clear case of eustress that builds our body and improves our survival ability.

This explanation of good and bad stressors and the principle of hormesis gives us a context to address the question: Why are people in Western culture today apparently so apathetic in the face of threats to its existence? While various forms of "bad" stress prevail, the impetus of "good" stress is simply not predominant. Food is abundant, and relatively speaking we have lived for decades without immediate violent threats from the outside. In terms of absolute survival, our existence is not under duress; and this results in extreme apathy, hedonism, and a lack of motivation. Subsequently, both low testosterone and low cortisol deter people from being willing to fight for their liberty and culture, particularly in the face of threats from the high-stress, high-cortisol Islamic culture of submission, violence, and aggression.

The lack of compelling eustress in a relatively well-off and comfortable culture might also account for the phenomenon of cultural evolution and the generation gap in the rise and fall of family fortunes. The saying "shirtsleeves to shirtsleeves in three generations" states the fact that most wealthy families lose their fortunes in three generations. Second generation wealth decreases in sixty-five percent of wealthy families, and by the third generation, that percent jumps to ninety-five:

As a rule, people who struggled up the proverbial rags-to-riches path are highly motivated, even obsessive. Lacking that drive, their children, are usually better at spending the accumulating money. Grandchildren are even bigger spendthrifts.

Offspring of the super wealthy often suffer from "affluenza," he says, crediting Jesse O'Neill, author of *The Golden Ghetto: The Psychology of Affluence*. They lack purpose, self-esteem and the ability to delay gratification.[17]

## Nature versus Nurture

When considering differences between generations, we are naturally led to the question of "nature versus nurture." This phrase encapsulates the long debate over the relative significance a person's innate characteristics play in shaping personality and leading to individual differences in behavior compared to one's life experiences. Nurture, of course, refers to behaviorism, while nature refers strictly to genetic inheritance.

The phrase was popularized by Francis Galton (1822-1911), the English scientist and founder of the eugenics movement who is associated with the study of behavioral genetics and the relation of heredity versus social environment of human character and development. Charles Darwin's book *On the Origin of Species* was a major influence on Galton and on the development of eugenics as a program to further the evolution of the human species as an experiment in animal breeding.

As presented in chapter 1, ideas pertaining to nature and nurture, a philosophy of the human mind, have altered throughout the centuries. A brief review is worthwhile here. The Enlightenment philosopher John Locke stated the "nurture" position, the concept of humans as blank slates who acquire their behavioral traits from their nurturing environment. This elevates the role of reason in framing our mind and understanding of the world and it carried through much of Western thinking until the nineteenth century, when Idealists like Hegel and Romanticists like Rousseau began favoring changeable emotion over reason. As testosterone declined into the 1930s, a new school of behaviorism emerged that determined complete dominance of the cultural environment on human development, elevating the emotions and feelings over reason and cognition, but disregarding the role of instincts.

In the period of the 1940s to 1960s, a renowned proponent of behaviorism was Ashley Montagu, who allowed no contribution from heredity:

> Man is man because he has no instincts, because everything he is and has become he has learned, acquired, from his culture... with the exception of the instinctoid reactions in infants to sudden withdrawals of support and to sudden loud noises, the human being is entirely instinctless.[18]

In the 1960s, however, a period of testosterone decline, Robert Ardrey published *African Genesis* (1961) and *Territorial Imperative: A Personal Inquiry into the Animal Origins of Property and Nations* (1966) and suggested that innate motivations drive the human pursuit of territoriality. This line of thought was further pursued by Desmond Morris in *The Naked Ape* (1967). As social mood continued to decline into the 1970s, Harvard biologist E. O. Wilson initiated additional attacks on both the blank slate hypothesis and behaviorism with a series of books that included *Sociobiology* (1975) and *On Human Nature* (1979). In his writing he elevated the role of animal instinct in structuring human social life.

Two decades later, Judith Rich Harris took yet another turn in explaining what forms us. She published *The Nurture Assumption: Why Children Turn Out the Way They Do* (1998), stating that peer group pressure is the primary social force that shapes our character. Her innovative concept was that it is not the family that shapes a child's psyche as a teenager and adolescent, but his or her social group. Her answer to why children turn out the way they do seemed to be that parents matter less than you think, and peers matter more. Group socialization theory hinges on the observation that during adolescence, growing children tend to spend more time with their peers than parents, whom they rebel against to build their own personal identities. In addition, studies in behavioral genetics discover that up to fifty percent of the personality difference between adults originate in genetic differences. This means that only a very small part of an adult's personality is influenced by the home environment conditioned by the parents.[19] Further supporting these ideas, the best-seller, *The Blank Slate: The Modern Denial of Human Nature,* was published by Steven Pinker in 2002. In his writing he denied both the blank slate and behaviorism as dogma. He attacked the idea of a "ghost in the machine"—the existence of a human soul that makes moral choices detached from biological drives.

*Dual inheritance theory* (DIT), also called *gene–culture coevolution* states that human behavior is produced by the interaction of two primary evolutionary forces: genetic evolution and cultural evolution. The mechanism is a continuous interaction of genes and culture in a feedback loop, in which cultural transformations can influence genetic selection and vice versa. Culture is defined as "socially learned behavior" that is transmitted between the social organisms through various social learning mechanisms, such as conformity bias. Cultural traits significantly influence the social environmental conditions in which genetic selection operates. An example of how genes and culture coevolve can be seen in adaptations to the rise of agriculture and dairy farming in human civilization over the past millennia. This has caused genetic selection in humans for traits that facilitate the digestion of starch and lactose, respectively.[20] Recent research also supports another manifestation of gene-culture coevolution: the notion that a combination of our ancestors' cultural experiences and subsequent genetic responses affect future generations. Trauma, for example, that exacerbates anxiety or substantive trials that build resiliency reveal their marks by intergenerational epigenetic inheritance:

> According to the new insights of behavioral epigenetics, traumatic experiences in our past, or in our recent

ancestors' past, leave molecular scars adhering to our DNA. Jews whose great-grandparents were chased from their Russian shtetls; Chinese whose grandparents lived through the ravages of the Cultural Revolution; young immigrants from Africa whose parents survived massacres; adults of every ethnicity who grew up with alcoholic or abusive parents — all carry with them more than just memories.[21]

Having presented this history of our philosophy of mind, we can observe that it follows cyclical trends probably driven by hormonal factors. Our mission as rational man should be to transcend our biology and to reach ever closer the godly-inspired ideal stated by Locke of a rational mind.

## The Generational Cycles

Authors Neil Howe and William Strauss have done much analysis on the characteristics of current generations, writing such titles as *Generations*, *The Fourth Turning*, and *Millenials Rising*. They posit that social mood exhibits itself in cycles of a "recurring and fixed order." Specifically they identify four twenty-year cycles, which, like seasons, are characterized by particular features. They call these seasons "turnings," beginning with a "High" that follows a "Crisis."[22]

In these First Turnings, individualism is preempted by institutions. "Society is confident about where it wants to go collectively, even if many feel stifled by the prevailing conformity." In a *Washington Post* article, Howe cites the post-War US in the late 1940s through early 1960s as well as the eras following the framing of the American constitution and following the Civil War. These were all times when the family was a stable institution and the society prospered industrially. The Second Turning is an "Awakening," as individuals seem compelled to assert more personal autonomy. They are seeking "higher principles and deeper values," a quest that leads them to criticize and even reject standing institutions of government and religion. Howe claims that this comes from a kind of progress fatigue when people "suddenly tire of all the social discipline." The revolutions of the 1960s and 70s certainly are examples of a Second Turning.[23]

It should be noted that while the form of Howe and Strauss's model aligns with my model of cyclical change, their analysis of the characteristics of this cycle in particular differs. What they extol as strength in individualism in a Second Turning, the Testosterone Hypothesis considers as decline into group-think resulting from testosterone decline. The deterioration of group cohesion caused by heightened social anxiety then naturally leads to their model's Third Turning, which is an "Unraveling." Here is how the Third and Fourth Turnings can be understood:

> The Third Turning is an "Unraveling," in many ways the opposite of the High. Institutions are weak and distrusted, while individualism is strong and flourishing. Third Turning decades such as the 1990s, the 1920s and the 1850s are notorious for their cynicism, bad manners and weak civic authority. Government typically shrinks, and speculative manias, when they occur, are delirious.
> Finally, the Fourth Turning is a "Crisis" period. This is when our institutional life is reconstructed from the ground up, always in response to a perceived threat to the nation's very survival. If history does not produce such an urgent threat, Fourth Turning leaders will invariably find one — and may even fabricate one — to mobilize collective action. Civic authority revives, and people and groups begin to pitch in as participants in a larger community. As these Promethean bursts of civic effort reach their resolution, Fourth Turnings refresh and redefine our national identity. The years 1945, 1865 and 1794 all capped eras constituting new "founding moments" in American history.[24]

Published in 1997, *The Fourth Turning* predicted aspects of the 2000 decade that indeed came to pass. The authors claimed that in or around the year 2005 a "Great Devaluation" would occur in world financial markets, and they correlated such a turn to the 1930s. Commenting on their prophesy in the 2017 article, their striking parallels certainly ring true:

> In the economy, both decades played out in the shadow of a global financial crash, and were characterized by slow and disappointing economic growth and chronic underemployment of labor and capital. Both saw tepid investment, deflation fears, growing inequality and the inability of central bankers to rekindle consumption.
>
> In geopolitics, we've witnessed the rise of isolationism, nationalism and right-wing populism across the globe. Geostrategist Ian Bremmer says we now live in a "G-Zero" world, where it's every nation for itself. This story echoes the 1930s, which witnessed the waning authority of great-power alliances and a new willingness by authoritarian regimes to act with terrifying impunity.
>
> In social trends, the two decades also show parallels: falling rates of fertility and homeownership, the rise of multi-generational households, the spread of localism and community identification, a dramatic decline in youth violence (a fact that apparently has eluded the president), and a blanding of pop youth culture. Above all, we sense a growing desire among voters around the world for leaders to assert greater authority and deliver deeds rather than process, results rather than abstractions.[25]

Howe sees Western culture careening toward a crisis or Fourth Turning, if not already in the midst of one. He sees the demise of liberal democracy and warns about the "creative destruction of public institutions" that will ultimately shift power from the old to the young by destroying what's deemed obsolete.[26] Quoting Lenin, he states, "In some decades, nothing happens; in some weeks, decades happen." History displays these crisis states every eighty years or so, according to Howe, measuring between Colonial America and its Revolution, the American Revolution and its Civil War, the Civil War and the Great Depression, the Depression and the financial crisis of 2008.[27]

Howe and Strauss's intriguing model echoes the writings of Oswald Spengler who wrote *The Decline of the West*, published between 1918 and 1923. Spengler was responding to the decline in social mood that led to the First World War. He also rejects the linear view of history as progress for a cyclic view of cultural evolution of the superorganism through different epochs. In that era he determined that the West was in its final season, or "winter time," of the Faustian epoch of civilization, which would only end in tragedy. Little could Spengler have known how the fears of his day would be reiterated within the course of the century following his predictions.

## Cultural Revolutions

Periods of testosterone decline, such as in the 1970s, lead to cultural revolutions led by the adolescent generation, whose growing personality and conflict with the older generation shapes the future of civilization. Sometimes the revolt has fomented among young people, such as the anti-war and sexual revolutions in the US during the 1960s and 70s; and sometimes students are used as the vanguards of revolution, their inherent discontent fueled by political agitators within universities or the press. Most often, it's a combination of these factors that fuels the fires of rebellion into cultural revolt. We need not look hard to see such cultural revolutions, be it the French Revolutions of the mid 1800s or the Russian Revolution of 1917. Young people serve as powerful mechanisms of significant historical changes.

The Cultural Revolution in China was a social and political movement from 1966 until 1973. Communist Chairman Mao Zedong was intent on purging any capitalist and bourgeois elements from the Chinese culture through a Marxist class struggle. The youth responded, forming Red Guard units, pursuing Mao's personality cult, and promoting the "back to the countryside" movement, with millions sacrificed for the motherland. The Nazi Party was also infamous in its systematic indoctrination of youth, creating for itself a generation of soldiers who were ready to fight and die for the Fatherland by the mid-1930s. The revolution in this case was against any entity or ideal that did not align with the promise of National Socialist dogma, including the "old ways" of prior generations and certainly the "other ways" of competing people groups:

> From the 1920s onwards, the Nazi Party targeted German youth as a special audience for its propaganda messages. These messages emphasized that the Party was a movement of youth: dynamic, resilient, forward-looking, and hopeful. Millions of German young people were won over to Nazism in the classroom and through extracurricular activities. In January 1933, the Hitler Youth had only 50,000 members, but by the end of the year this figure had increased to more than 2 million. By 1936 membership in the Hitler Youth increased to 5.4 million before it became mandatory in 1939. The German authorities then prohibited or dissolved competing youth organizations.[28]

Another historical event propelled by the young generation is recorded in the first book of Kings in the Bible. The new King Rehoboam listened to the advice of the young men who were rebelling against the older generation, promoting a socialist revolution but telling him to increase the tax burden on the nations:

> And the young men that were grown up with him spake unto him, saying, "Thus shalt thou speak unto this people that spake unto thee, saying, 'Thy father made our yoke heavy, but make thou it lighter unto us;' thus shalt thou say unto them, 'My little finger shall be thicker than my father's loins.
>
> 'And now whereas my father did lade you with a heavy yoke, I will add to your yoke: my father hath chastised you with whips, but I will chastise you with scorpions.'"
> (1 Kings 12:10,11)

The ensuing tax revolt and other events eventually caused the Kingdom of Israel to split into two kingdoms in 930 BCE.

In conclusion, we humans as biological organisms, are designed to evolve in order to adapt to changing environmental conditions. Though we might not realize it, the profound social transitions in our cultural evolution on a time scale of a few human generations are triggered by biological mechanisms that emerged over millions of years of evolution. With the rise of Homo sapiens, we gained the faculty of reason, but it did not replaced our biological drives, only augmented them with greater abilities. However, employing our great capacity for learning we will continue in the next chapters to study how the origins of sex and social behavior have come to shape our minds and form the generation gap, regulated by hormonal cycles.

# 4

## THE SOCIAL ANIMAL

*No man is an island.*  — John Donne

*There's no such thing as society. There are individual men and women and there are families. And no government can do anything except through people, and people must look after themselves first. It is our duty to look after ourselves and then, also, to look after our neighbors.*[1]
— Margaret Thatcher

These quotes from John Donne and Margaret Thatcher perfectly capture the tension of human existence: We are indeed individuals, and we are indeed members of groups— two states of being played out in the perpetual conflict of Western politics, which is individualism versus collectivism. How do we organize ourselves socially? Why do we seem to swing between apparent extremes in social organization? The previous chapter described the pendulum swings as seen between generations, and certainly models like Howe and Strauss' "Turnings" portray an even longer cycle of these societal alterations. However, there is much evidence that our species is capable of evolving our social organization to a mode between the two extremes of utter individualism and complete collectivism. Humanity is a homo-duplex species, and hormonal cycles that turn certain genes on or off allow for epigenetic evolution of our cultures.

## Truly Social Animals

Throughout history we have investigated other creatures, comparing and contrasting their "social" structures or behaviors with our own. The wise King Solomon exhorted in his proverbs to "Go to the ant, you sluggard; consider its ways and be wise! It has no commander, no overseer or ruler, yet it stores its provisions in summer and gathers its food at harvest" (Prov. 6:6-8 NIV). The thread of this inclination runs through all cultures, and in our Western lineage, we see it in Aesop's Fables, popular medieval allegories, Darwin's biology, and the sociobiology of E.O. Wilson, to name just a few examples. In this chapter, we will look specifically at academic study of *eusocial* animals. These are animal groups that organize themselves at the highest levels of sociality or cooperation.

There are at least a few reasons to focus on these studies: As evidenced in this book, there is a strong human propensity to move toward those communal systems of sexual organization when testosterone declines; declining social mood affects the very perceptions that motivate thought leaders, like academics, to favor eusociality. The influence of social mood on academia was presented in some detail in chapter 1, where I noted that contemporary sociobiology has come to see this model of eusociality as most virtuous, thereby worthy of comprehending, if not emulating. Therefore, it behooves us to understand what eusocial creatures act like, the role hormones play in certain facets of their societal organization, and how scientists identify humans with them.

*Eusociality* means truly social. It is the peak level of sociality in animal group organization, characterized by cooperation in caring for the brood, for group division of labor into a caste system of specialized reproductive and non-reproductive members, and for the overlapping of generations within the colony. One distinguishing aspect of eusociality is that some individuals lose the ability to perform tasks, which in turn are relegated to other members of the social group, for example, the ability to reproduce. Eusociality exists in some insects, crustaceans, and even mammals. In mammals, two species of mole-rats are eusocial; among crustaceans, certain varieties of shrimp exhibit eusociality. It is most obvious in the study of *Hymenoptera*, which includes ants, bees, and wasps, and also in termites. The caste differentiation in a colony is between the reproductive members, the queens and males that are solely responsible for fertility, and the worker and soldier castes that labor to create adequate living conditions to raise the brood. Eusocial creatures comprise a superorganism: a group of organisms in the same species that interact in synergy to create a social unit. The collective acts in concert to produce the emergent phenomenon of a greater whole than the sum of its parts. Darwin determined that the unit of evolutionary selection in a eusocial group is the group as a whole, not each individual creature.

The term *eusociality* was first introduced by Suzanne Batra in 1966 to describe the cooperative nesting behavior in one species of bees, where an individual starts the colony of related helpers. Academic understanding of eusocial animals is exemplified in the past forty years of E.O. Wilson's writings. In the period of testosterone decline of the 1960s and 70s, when the sexual revolution changed human sexual behavior, Wilson published *The Insect Societies* (1971), extending eusociality to other social insects. His landmark 1975 book, *Sociobiology: The New Synthesis*, proposed correlations between eusocial animals and human behavior, as did his 1990 publication, *Ants*. In 2012, in his book, *The Social Conquest of the Earth*, he suggests the humans are a species of eusocial ape.

Wilson's contemporaries in the 1970s found his ideas abhorrent. Such genetic determinism, they argued, resonated with racism and, even worse, fascism. In Wilson's own words:

> Sociobiology then came under attack by critics all over the place because its use in studying human behaviour. It was regarded as biological determinism which was not acceptable for the social sciences. Any idea that human

behavior of any kind had a biological basis was not acceptable in the seventies. And then there were Marxist critics like [Stephen J.] Gould and [Richard] Lewontin who felt that it was injurious to the progress of human beings toward a socialist society, which they considered the most just and inevitable society. You won't get Gould admit that today, but that was how he talked in those days![2]

By the 1990s, however, sociobiology was gospel. Ideas in evolutionary psychology were also changing, as described in chapter 1, and the scientific world was reducing mankind to instinctual, animalistic, group behavior on par with the social insects, such as ants and bees. The hormonal cycles that cause epigenetic evolution of reproductive strategies in the human species had, in the short period of a few decades, also altered our scientific understanding of our own sexual organization. What scientists like Wilson were missing is the fact that humans are much more complex than animals. We experience phase transitions between the communist and the individualist models according to our societal testosterone levels. This is in contrast to eusocial species that have a fixed, immutable, communist social model of organization.

Inherent in the sociobiological models of evolution is the inclination to idealize what seems to be a pure form of a desirable trait. Therefore, if *apparent* "altruism" in eusocial animals can be isolated and understood, then in a utopian mindset, perhaps it gives us critical information to move closer to promoting this trait in human culture. In the writings of these scientists, one cannot help but recognize a kind of admiration for the eusocial species, as if their "altruistic" behaviors that produce such efficiency are to be envied. The danger in this sentiment, however, is losing our higher, rational perspective. If we anthropomorphize the ant or the mole rat's cooperative behavior as some kind of noble virtue, we risk forgetting that the creature is *genetically compelled* to function as a unit with its group. In other words, if we look to the ant and superimpose "altruism," or "selfless acts," we make a category mistake. We attribute moral agency to a creature that is merely mechanistic. And, on the other hand, if we detach human volition from our moral choices by seeing humans through the lens of mere genetic mechanism, we belie the crowning distinction of rational man—free will.

## The Evolution of Eusociality

A fundamental question for biologists, dating back to Darwin, is how natural selection chose "selfless" behavior that would seem to contravene survival of the fittest. With eusocial animals it turns out that the survival programmed into the individual is that of the group. In some ants and bees, that colony- or hive-level selection favors disease resistance. Also, nest temperature stability is critical to these species, and they have evolved to promote that. A 2010 article published in *Nature* distills the stages of eusocial evolution into the following, very general stages: 1) group formation; 2) tight formation of the group made possible by "a valuable and defensible nest;" and 3) because of that "durable nest," group persistence is sustained. To that end, mutations "silence dispersal behavior," minimizing disruptions to the colony.[3] For many years, the leading theory explaining eusocial behavior was "kinship" or "inclusive fitness." These compelling ideas generated from the recognition that ants of certain species were up to seventy-five percent related. Therefore, the theory posited, high degrees of relatedness motivated creatures to ensure each other's survival. However, further research found far too many other eusocial species that were not highly related, leaving current researchers to this kind of conclusion:

> [C]onsidering its position for four decades as the dominant paradigm in the theoretical study of eusociality, the production of inclusive fitness theory must be considered meager. During the same period, in contrast, empirical research on eusocial organisms has flourished, revealing the rich details of caste, communication, colony life cycles, and other phenomena at both the individual-selection and colony-selection

levels. In some cases social behavior has been causally linked through all the levels of biological organization from molecule to ecosystem. Almost none of this progress has been stimulated or advanced by inclusive fitness theory, which has evolved into an abstract enterprise largely on its own.[4]

Among the flourishing empirical research on the social insects, studies reveal the significant role hormones play in social organization. The hormone dopamine makes genetic pathways that drive both dominance and reproductive activity in ants. If a colony's queen perishes, dopamine levels are two to three times higher in ants that win ritual domination battles to become queens-in-waiting, called *gamergates*. The dopamine changes in these worker-queens correlate with greater activity in their ovaries, increasing their reproductive potential; and conversely, dopamine lowered quickly during these dominance tournaments in the ants that lost, thus diminishing ovary activity.[5] The traits that lead different creatures to do different tasks in various castes of eusocial insects is also apparently regulated by hormones such as serotonin and tyramine.[6] The larvae maintain an ability to differentiate into either queens or workers. Once a developmental progression is "committed to," the creatures lose what is called *phenotypic plasticity.*[7] The subsequent caste identification cannot easily be altered. Researchers see parallels between this process and that of cell differentiation in mammals and are developing models to better understand the mechanism of plasticity with an eye to a better framework for biological complexity.[8]

While E.O. Wilson came to identify humans as eusocial, he acknowledged that the kind of communal cooperation that evolved in fully eusocial creatures did not evolve in humans, specifically because of our reproductive independence. He writes:

> What I like to say is that Karl Marx was right, socialism works, it is just that he had the wrong species. Why doesn't it work in humans? Because we have reproductive independence, and we get maximum Darwinian fitness by looking after our own survival and having our own

offspring. The great success of the social insects is that the success of the individual genes are invested in the success of the colony as a whole, and especially in the reproduction of the queen, and thus through her the reproduction of new colonies.

We now understand quite well why most species of social insects have sterile workers, and therefore can have communist-like systems in which the colony is all, the individual is only a part of the colony, and the success of the whole community is what counts far above the success of the individual. The behavior of the individual social insect evolved with reference to what it contributes to the community, whereas the genetic fitness of a human being depends on how well it can individually use the society. We have become insect-like only by extreme contractual arrangements.[9]

It's not hard to identify a certain intonation in Wilson's words. The social insect "contributes" to the group, while a person seeks to "use" the group. The insect structure is his reference point in the final sentence: our only way to be "insect-like"—as if that's what we attain to—is "extreme" social contracts. The premise of his description seems to be the desirability of communal social organization, with the lament that man can only get there via considerable effort.

### The importance of the male role

The individuality that defines a strong human society is a result of males fulfilling a primary and continuously active role in the sexual organization of society. This is not a popular reality to contemporary academics like Wilson. In human society, the male takes an active, dominant, and leading role in forming the family unit, laboring to provide for the young and his female counterpart, and building social bonds with other member of society. Compare this to ants, in which only females compose the nest, the male's singular role being to fertilize the queen that preserves his semen for a long period of time in order to fertilize her eggs during her entire life cycle.

# Sex Wars

In contrast to ants, whose sexual organization is a fixed structure, with the queen taking the sole reproductive role within the caste system, humans exhibit much more diversity in a variety of social systems. The degree of independence in a society aligns with the degree that male independence is valued in the social structure. The rhythm of human civilization's history is driven by recurring cycles of the hormone that literally shapes the social structure of human cultures—testosterone. A highly masculine culture, such as the model presented in the Hebrew Bible, will idealize masculine leadership for independence through the formation of strong family units in a patriarchal social structure. The individual will be deemed as the center of human existence and society as an emergent form of social network through the consensual and contractual agreement between free men. This is not a condition to be disdained, rather a creative and fascinating capacity of our rational minds. The entire culture, as we see in ancient Israel, will view itself from a dominance position, seeking to separate itself from foreign people, the pagans or idol worshipers, as they are referred to in the Bible. However, as testosterone levels drop, the social structure will evolve accordingly and the culture will gradually collectivize. It will lose its degree of independence from within and finally succumb to foreign barbarian forces that had once seemed inferior taking it over from without, as we have witnessed many times in the history of human civilization.

## The family structure

The word *husband* originates from the Old Norse word hūsbōndi, meaning "master of a house," and in Hebrew the word is baal (בעל), meaning owner. This patriarchal social structure is rooted in the evolution of sexual reproductive strategies in which males compete for the females of the species and seek to fertilize a female. In the human species in particular, the parental investment is huge, as the development process of a child takes more than two decades. This is due to the large brain size of humans, which requires an extended period of learning and adaptation to its environment for the benefits of producing an intelligent individual. We are born with thirty percent or less of our brain size, the advantage of which is the resultant experiential learning with our environment, as opposed to the brain growing in utero. For the human fetus' brain to develop just ten percent more during gestation, it would take a year-and-a-half to almost two years.[10]

This long developmental period, as presented in chapter 2, might account for the evolution of monogamy in humans. Researchers are not definitive about why humans incline toward monogamy, although biparental care and "mate-guarding strategy" throughout a long gestation and nursing period are indicated along with the more vulnerable offspring.[11] Most particularly, however, researchers who collected data from 230 primate species found that the potential killing of male infants correlates most highly with monogamy:

> It is only the presence of infanticide that reliably increases the probability of a shift to social monogamy...The origin of social monogamy in primates is best explained by long lactation periods..., making primate infants particularly vulnerable to infanticidal males... [B]iparental care shortens relative lactation length, thereby reducing infanticide risk and increasing reproductive rates. These phylogenetic analyses support a key role for infanticide in the social evolution of primates, and potentially, humans.[12]

The significance of monogamous relationships in the human species and some primates can be contrasted with our chimpanzee relatives, who raise their young communally in a tribe. The alpha male in a chimpanzee band fertilizes most of the females, who also have sex with other males when he is not watching, so the offspring are raised by communal effort. In humans, direct parental investment and care of the father for his offspring within the family structure leads to greater effort and resources given for the benefit of the child.

Regarding how different species parent their young, ecologists have a theory called the *r/K* Selection Theory that presents two specific ways that creatures reproduce. With the *r*-selection strategy, species like bacteria, insects, grasses, and rodents reproduce abundantly. Their rate ($r$) of offspring production is high, and they tend to exist in spacious ecological niches without a lot of competition. Species that are $r$-selected produce numerous offspring, yet each organism has a low survival probability. The other type is the $K$-selected species. These include larger animals with longer parental-care commitments and lifespans, such as whales, elephants, and humans. Identifying traits for these species include their production of fewer offspring and their choice of crowded ecosystems that require competitiveness for survival. These offspring have a much higher long-term survival rate than $r$-selected offspring.[13]

Applying this model to human behavior, an author who uses the pseudonym "Anonymous Conservative" published *The Evolutionary Psychology behind Politics: How Conservatism and Liberalism Evolved within Humans* (2014) applying the biological motivations in the $r/K$ selection theory to explain the sociopolitical divide across the political spectrum. The author attributes characteristics of more liberal people to the "$r$-strategy," whereby they are "averse to all peer on peer competition, embrace promiscuity, embrace single parenting, and support early onset sexual activity in youth...and the sexualization of cultural influences."[14] These are all supported by the liberal cultural philosophy that simultaneously rejects "individual Darwinian competitions such as capitalism and self-defense with firearms, as well as group competitions such as war."[15] $K$-strategy, on the other hand, supports competition like capitalism and war, individual rights to self-protect with firearms, and tends to favor abstinence until monogamy, as well as two-parent families "with an emphasis upon 'family values,' and children being shielded from any sexualized stimuli until later in life."[16] Although this theory holds some merit in explaining the differences between the dominant, masculine-type group behaviors and the submissive feminine group behavior, it fails to encompass the great complexity of human social interactions and how it is mediated by sex and social status hormones.

# 4 • THE SOCIAL ANIMAL

## Gene-Culture Coevolution

The phenomenon of cooperation in humanity's superorganism is even more complex than among eusocial animals. Gene-culture coevolution is interwoven so intricately that we might only have begun to comprehend what is taking place. Social scientists attribute most of our "other-regarding" traits to the evolution of this dynamic:

> Gene–culture coevolution is responsible for the salience of such other-regarding values as a taste for cooperation, fairness and retribution, the capacity to empathize, and the ability to value such character virtues as honesty, hard work, piety and loyalty.[17]

Moreover, they consider it to be pivotal in the flourishing of our species in the milieu of our early hunter-gatherer communities. Thereby we evolved "the capacity of unrelated, or only loosely related, individuals to cooperate in relatively large egalitarian groups in hunting and territorial acquisition and defense."[18] Not only did the human race flourish, it has survived and become the dominant species on the planet. According to contemporary thinkers, the complex system of gene-culture coevolution is arguably responsible for that great success.

It was Richard Dawkins in *The Selfish Gene* (1976) who is credited with developing the theory that cultural memes function as instruments of epigenetic transmission, in other words, that culture mutates and evolves like genes do. Further, the micro realm influences the macro and vice versa. The study of our evolved methods of cooperation through religion or political groupings is a study, in part, of developmental moral psychology—how valued norms are transmitted and internalized from generation to generation, people group to people group.[19] And, at its core, then, it is also the study of hormones.

A manifestation of gene-culture coevolution of particular interest to this writing is the human social inclination toward phase transition. The driving mechanism of phase transitions is hormonal fluctuations, an impetus powerful enough over time to shift entire cultures from one set of normative morality, accepted religion practices, and preferred political structures to another set. Even if not all individuals agree with such alterations, they get swept along in the waves of change and learn ways to adapt, or else perish.

The endocrine forces that affect changes in human social organization are, of course, also seen in animals. Serotonin is an ancient hormone that drives social status and behavior in animals, such as insects as well as humans. Certain kinds of grasshoppers seem to have the same biological adaptation that humans do that allows them to transform from anti-social animals into cohesive group formations. The desert locusts functions as either loners or gregarious social creatures. In periods of drought, locusts live not merely as individuals, but in apparent aversion to socialization. Their mode shifts when rain leads to abundant food sources and breeding ensues. Inevitably, the rain stops, vegetation is widely consumed, and the locusts must congregate in smaller areas of available food. The crowding itself flips the serotonin switch and turns on the gregarious phase that creates the enormous, destructive swarms humans dread.

> This crowding triggers a dramatic and rapid change in the locusts' behaviour: they become very mobile and they actively seek the company of other locusts. This new behaviour keeps the crowd together while the insects acquire distinctly different colours and large muscles that equip them for prolonged flights in swarms.[20]

Recognizing this pattern raises the question why the locusts need to transition between the solitary and gregarious phases at all. The simple answer is cannibalism. While a swarm aids migration by deterring predators, it stresses social relations, so to speak. According to Mike Anstey of Oxford University, "Switching to the gregarious phase is really the lesser of two evils—individuals will certainly die of starvation if they do not migrate and are only quite likely to die by migrating."[21]

In addition to the eusocial animals earlier discussed, group behavior in other species is also regulated by hormones. Studying migration patterns, biologists have isolated cortisol and testosterone as hormones that increase prior to migration in sturgeon, for example. This biochemistry is seen in other fish and birds, especially prior to spring migration. Fall migration seems to be reliant on changes in thyroid hormone concentrations. [22] As noted above with the locusts, muscularity changes allow for greater aerobic capacity as animals ready for migration, and thyroid hormones contribute to those changes in certain geese species.[23] Such annual migrations show clear patterns, geographically and biochemically. Likewise, herd behavior and the complex systems of synchronized group movement in swarms, schools of fish, and flocks present fascinating phenomena that we are only beginning to understand.

When we turn our attention to human migration, we face even more complexity. At a very basic level, though, humans migrate for two predominant reasons: to find food and to flee conflict. For that matter, food scarcity might be the result of conflict, as well as catastrophic weather events and dangerous animal predators. In any of these cases, stress hormones are elevated, serving to compel the move and to cohere the group. The great migration we now witness of Muslims from the war-torn regions in the Middle East and Africa into Western Europe are driven by the same stress hormones that push the sturgeon to swim upstream.

Group coherence operates in many ways other than migration. Take for example how groups congregate for rallies and protests and "pride" parades. With some exceptions, these gatherings are most often seen when the group detects or experiences opposition. Serotonin is the status and pride hormone. Human males synthesize serotonin at a rate that is fifty percent higher than females, which researchers believe accounts for the lower rate of depression in men.[24] Yet, when serotonin levels drop and stress levels rise with higher cortisol, the human brain becomes depressed and seeks to raise its social mood by elevating its social status. Serotonin has also been linked with sexual preferences. In 2012, Peking University researchers changed the sexual inclinations of female mice by changing their serotonin receptors. When levels were diminished, the female mice preferred to sniff and mount other females. In concluding the study, the scientists assert, "Our results indicate that serotonin controls sexual preference."[25]

It is not surprising then that the homosexual community has a long-standing tradition of gathering in pride parades. In social events like gay pride parades and the enormous marches seen in communist Russia and China, individuals tend to lose their independence and faculty of reason in surrender to ancient animal instinct by forming cohesive groups that march together and promote unity. In addition to unity, the spectacle is decidedly *against* perceived enemies. This could mean the patriarchy for feminists, the capitalists for Marxists, the Jews for the Nazis, and infidels for the Islamic jihadists. Furthermore, this phase transition in human social behavior could be the primary engine of our major historical events, such as revolutions and wars. In contrast, in a high-serotonin culture that is associated with the masculine brain type, individuals tend to be happier with their lives, seeking independence and liberty, rather than group conformity and obedience.

Group coherence is, of course, not only episodic, operating during events that take place for a limited time like a migration period or a political rally. The principles underpinning any specific group connect the individuals over time. Chapter 1 reviewed the ideas about evolved cooperation in religion and politics, including the work of Jonathan Haidt. Haidt assesses the group dynamics of shared intentionality that led to codifying ethics and creating religion. Groups strongly cohered by religion, he argues, survived longer than those without shared religion and moral systems. Religion, therefore, is a survival adaptation.

Haidt aims for a nuanced understanding of group moral development that accounts for the ideas of other thinkers in his field. Richard Dawkins rejects group selection and understands social cohesion as essential driven by the selfish pragmatism of individuals. David Sloan Wilson (*Darwin's Cathedral*, 2002) proffers the negative aspect of religion—the stick as opposed to the carrot—that motivates cohesion with threats of punishment or shunning and builds trust and emotional connection over and against other groups. What Haidt proposes is that we function simultaneously as individuals adhering to social contracts that diminish self-centered behavior *and* as a hive entity that prospers by its survival traits. Beginning with his definition of morality, he describes these parallel modes:

> Moral systems are interlocking sets of values, practices, institutions, and evolved psychological mechanisms that work together to suppress or regulate selfishness and make social life possible.
>
> The contractual approach takes the individual as the fundamental unit of value. The fundamental problem of social life is that individuals often hurt each other, and so we create implicit social contracts and explicit laws to foster a fair, free, and safe society in which individuals can pursue their interests and develop themselves and their relationships as they choose.
>
> The beehive approach, in contrast, takes the group and its territory as fundamental sources of value. Individual bees are born and die by the thousands, but the hive lives for a long time, and each individual has a role to play in fostering its success. The two fundamental problems of social life are attacks from outside and subversion from within. Either one can lead to the death of the hive, so all must pull together, do their duty, and be willing to make sacrifices for the group.[26]

A social psychologist, Haidt pays much attention to current trends in group behavior. Not surprising to readers of this book, he cites an increase in "partisan antipathy" over the past two decades of American life—decades of testosterone decline. Polls of Democrats and Republicans that measured opinions between the parties show that the "very unfavorable" reporting about the other party more than doubled in that time. Democrat participants went from sixteen to thirty-eight percent "very unfavorable" opinions regarding Republicans between 1994 and 2014. Republicans went from seventeen percent to forty-three percent in their "very unfavorable" reporting about Democrats.[27] More stark political polarization indicative of tighter group cohesion correlates with all the other symptoms of low social mood in the recent downturn of society.

To end this chapter where we began, we return to a further consideration of eusocial tendencies in human culture throughout history. Critics of E.O. Wilson's assertion that humans are one of the few eusocial species on earth do make good points. They do not see people abdicating reproductive capability, for example, which is one of the hallmarks of eusocial societies. While this might be true in the case of high-testosterone cultures like the West has been over the last few centuries, it has not been universally true. As we have discussed before, the terms eunuch or castrati refer to people who serve some very particular functions in human societies by intentionally being denied their reproductive capacity. The Bible refers to the castrated servants in the courts of the kings in pagan societies, calling them "saris" in Hebrew. In in the peak-testosterone culture of the Hebrew Bible, individual fertility was blessed and castration was strictly forbidden. Furthermore, it could also be argued that cultures that support and even demand widespread abortion limit the reproductive abilities of women. These tend to be communist-organized systems like Russia and China and the contemporary societies of the West that are swiftly descending into cultures that restrict individualism. Even if feminism has promoted abortion as an individual "right" for a woman, it's a destructive act to women's bodies and psyches that only the collectivist *zeitgeist* has promulgated.

These same low-testosterone socialist ideologies that advocate methods of reducing birth rates have historically been susceptible to predictions of over population. In statist, progressive regimes, declines in conception rates are therefore desirable, even preferable. When there is a perceived lack of resources due to low levels of growth hormones, the group suppresses reproduction. This was true in ancient Greece where philosophers like Aristotle called for forced emigration, infanticide, and abortion if population in the city-states threatened to exceed what they considered optimal numbers for efficient governance.[28] Already in that culture, the acceptance and proliferation of homosexuality indicated decreased testosterone, and the correlation can be made to the increase of homosexuality along with abortion as well as fears about population sizes in current Western culture as testosterone levels have decreased.

Academic psychology bears out the trend that population concerns heighten as testosterone lowers. In the 1930s as well as the 1960s and 70s, overpopulation was given serious—if not at times panicked—attention. Whether it was Paul Ehrlich's prediction of mass starvation and death in his 1968 book *The Population Bomb*, or the rodent research of John Calhoun that investigated the psychological dilemmas of over-crowding, pessimism abounded. Calhoun's dire predictions about how urban over-crowding would lead to mental illness and moral decay were countered by research that clarified the dynamic. It was not the population density in and of itself that caused higher levels of aggression and stress, it was the amount of unwanted social interaction. Therefore, in rats and humans alike, the better that creatures can modulate their private–public exposure, the calmer they are. There is learned behavior to manage one's environment, and people in densely populated environments simply require attention to those skills.[29]

Adolf Hitler's entire social engineering agenda and environmentalism was motivated by his fear that the Aryan race would not have enough resources to prosper. The ideology of *lebensraum* reached its pinnacle under Hitler's regime as the Nazis fought to conquer greater living space for their people. Farming and a connection to the land was essential to Nazi ideology and drove them to purge the crowded urban areas particularly associated with Jewish communities. "Blood and Soil" had long been a principle of German nationalist romanticism, whereby the rural life was exalted as the haven for true German ideals. In 1930, the phrase was promoted through the writings of Richard Walther Darré who published *Neuadel aus Blut und Boden* [A New Nobility from Blood and Soil]. Darré's book spurred on the rural German population with his call to return to the land, to breed pure-blooded Gemans, and to purge the culture from other races—notably the Jews.[30] The fervor of this mythic ecology has been labeled "ecofascism," and its description by historian Michael Zimmerma echoes of eusociality: "[A] totalitarian government that requires individuals to sacrifice their interests to the well-being and glory of the 'land'..." Land, in this case, connotes the nation and race as well as physical territory.[31] The *Wikipedia* entry illuminates this idea well:

> Peasants were the Nazi cultural heroes, who held charge of German racial stock and German history—as when a memorial of a medieval peasant uprising was the occasion for a speech by Darré praising them as force and purifier of German history. This would also lead them to understand the natural order better, and, in the end, only the man who worked the land really possessed it. Urban culture was decried as a weakness, "asphalt culture," that only the Führer's will could eliminate — sometimes, as a code for Jewish influence.[32]

Other historical manifestations that align with eusocial behaviors can be seen in the creation of strict castes in various societies. In the low testosterone, statist culture of ancient Greece, Plato's tripartite theory of soul determined that biologically, society divides into major castes. These comprised the ruling class of philosopher kings, soldiers responsible for enforcing order as commanded by the kings, and at the bottom, the producers. Similarly, in India, Hinduism demands a hereditary caste system based on birth, dividing the population into five distinct groups: At the top are priests, called Brahmin, next are the ruling and military class, followed by the merchants and farmers, then peasants, and finally the untouchables at the bottom of the hierarchy. The three top classes are of Aryan racial descent, while the two bottom classes are non-Aryans. The Aryan migration theory determines that around the second millennium BCE, the Aryan people ("Aryan" meaning noble) from the region around the Black Sea spread into both Europe to the West and India to the East on the Eurasian continent. The Nazis adopted the idea of the Aryan race from Hindu mysticism and were also obsessed with genetic purity and their heredity as the super race.

# 4 • THE SOCIAL ANIMAL

The war between the Aryans and the Jews might actually be a millennial-old struggle driven by hormonal cycles that led to replicating genes battling for social dominance and submission. As I described in *The Objective Bible*, the Bible tells us that Abraham was born as Abram, which is quite similar to the Hindu priest called Braham. However, God changed his name to Abraham as the father of the Hebrew people. Abraham originated from Ur in ancient Sumer. The period of Sumer's fall, around the twentieth century BCE, is the same period in which the Aryan migration, probably from Persia, invaded India and created the caste system that placed them at the top of the Hindu dominance hierarchy with the indigenous population as subordinates.

One plausibility is that when testosterone increased again, the ancient Hebrews rebelled against the pagan system of social repression with what Camille Paglia calls the male declaration of independence from the ancient mother cults, articulated in the philosophy of biblical monotheism. This power struggle for social dominance versus independence was glamorized by Friedrich Nietzsche in his book *Zarathustra*. The name refers to a prophet who is to be the *superman* character. Zarathustra is a similar name to the prophet Zoroaster of the ancient Persian religion. The Persians are considered an Aryan race, and the king changed the name of the nation to Iran in 1935, during the period of Hitler's reign in Germany, to declare their strong ties of loyalty to the Aryan race and mythology.

The German, Persians, and ruling caste of Hinduism belong to the Aryan race, and this explains why the Nazis were enamored with eastern mysticism, practiced yoga, and chose the swastika, an ancient Hindu symbol, as the source of their pagan cult of submission. Their goal to exterminate the Jews as a race and not only as a religion, attests to the epigenetic roots of group conflict in the struggle for survival. The East Asian Hindus, having lower testosterone and dopamine levels, have never mastered the will to revolt against the Aryan system of oppression. The ancient Hebrews, however, given their masculine culture of dominance and independence, led to the rise of the modern Western civilization with its values of life, liberty, and individual rights. This battle of hormones and genetic tendencies in our species continues to play itself out in our contemporary culture and political wars.

# 5

---

# THE ORIGINS OF SEX AND DEATH: SYMBIOGENESIS

*The certainty of death was absent at the origin of life. Unlike humans and other mammals, many organisms do not age and die. The process of programmed, inevitable death evolved only after our symbiotic microbial ancestors, some two billion years ago, became sexual individuals. Sex began when unfavorable seasonal changes in the environment caused our protoctist predecessors to engage in attempts at cannibalism that were only partially successful. The result was a monster bearing the cells and genes of at least two individuals (as does the fertilized egg today).[1]*
— Lynn Margulis

In order to understand the origins of our cultural evolution and sex wars we have to explore the origins of complex life and sex. Only by understanding the emergence of complexity and sexual reproduction can we reach the big picture view of human evolution and life history.

At the origin of life there existed only primitive cells, each lacking a nucleus, with its DNA floating inside the cytoplasm, the cell's inner materials and fluids. These are called *prokaryote* organisms, and they include bacteria and other micro-organisms. According to the theory of Symbiogenesis, promulgated in the 1970s by eminent biologist Lynn Margulis and accepted today by the scientific establishment, two different prokaryote cells merged together to form the complex *eukaryotic* cell. In this organism, the DNA is contained in the cell nucleus as a separate organelle of the cell. The mitochondria, the energy producing units of this "modern" cell, were previously independent bacteria that were engulfed and merged into the complex cell in a symbiotic relationship that led to the origin of complex, multicellular life.

The primitive, prokaryotic life forms do not have the sexual life cycle of birth, sex, and death that complex organisms do; they simply transfer genes between themselves through a kind of swap called a horizontal gene transfer. In addition, aging and death is not programmed into their DNA, and they reproduce by cloning themselves asexually. With the introduction of complex, multicellular life, the life cycle we know in animals emerged to regulate the evolution of complexity. But how did this two-party reproduction come about? And further, why did it necessitate death?

In the article *Did Sex Emerge from Cannibalism? Sex, Death and Kefir* (1994), Lynn Margulis suggests that the original event leading to sexual procreation was an accident— "a desperate strategy for survival" gone wrong.[2] Changes in the environment of ancient protoctists (i.e., asexual reproducing organisms, the ancestors to fungi, plants, and animals) led to "attempts at cannibalism that were only partially successful." She calls the outcome of this cannibalistic desperation "a monster" that embodied the cells and genes of both organisms.[3] These new organisms became the sexual beings that mated and propagated by fertilization and cell fusion and in which programmed death developed.

Similarly to other scientists, Margulis' cannibal theory answers the question of why sex evolved: because of stress. As discussed in chapter 2, sexual reproduction serves to re-shuffle genes, allowing for variation at the same time that it filters disadvantageous mutations. Prokaryotic organisms lacked genetic diversity, and in times of particular environmental stress (e.g., starvation, temperature extremes, etc.), certain varieties of them became capable of optional interbreeding or *outcrossing*. In research with *Caenorhaboditis elegans*, a kind of worm, induced stress has shown the species' ability to change from self-fertilizing to predominantly outcrossing.[4] Likewise, such a shift in reproductive strategy and gene expression is seen in algae and has been understood as required for DNA repair.[5] A necessary component of survival is DNA repair; therefore, significantly stressed asexual organisms might well have evolved for this purpose, as well as variation. Researchers first hypothesized this in the early 1990s when seeing how over-heated algae switched on the organism's "sex-inducing gene," and cells began to interbreed with one another. In addition to the ability to switch reproductive modes, this phenomenon further indicated that the outcrossing process itself includes a DNA repair mechanism.[6]

Once organisms evolved that were fully multi-cellular and breeding by fertilization, the specialization of cells evolved: Germ cells house reproductive DNA and the repair capability, undifferentiated stem cells become specialized and can also create more stem cells, and already differentiated somatic cells specialize in all manner of biological tasks. Professor William Clark, author of *Sex and the Origins of Death* (1998), asserts that the programming of cellular death was an adaptation to protect organisms from the potentially faulty DNA of somatic cells.[7] It is necessary to recall that single-cell creatures like bacteria are not encoded with programmed self-destruction. They are, in Lynn Margulis' words, "eternally young." Clark presents cell death and death of the organism as the necessary mechanism for the benefits of reproduction to be optimized.[8] Insomuch as DNA is compelled to replicate, the old must give way to the new.

A more specific mechanism of cellular evolution, and thereby sexual evolution, can be seen in mitochondria. These intracellular packets, so to speak, contain their own DNA and produce energy for each cell. Their DNA is found in the female gamete, the egg, and is therefore passed through the female line of a species. Biochemist Nick Lane of University College London, uses the metaphor of "enslavement," proposing that these cellular engines were once single-celled bacteria subsequently captured within a larger cell.[9] His metaphor echoes Margulis' cannibalism, displaying biologists' attempts to describe the radical processes that have fueled evolution. And if these terms connote violence in evolutionary development, the outcome of such cannibalism or enslavement is the ultimate violence—death. In the economy of cell production where vastly more cells die than survive over time, energy-producing mitochondria generate free radicals that cause mutations at a higher rate than the cell's separate nucleus and thereby hasten cellular deterioration that leads to disease and death.

Evolutionary history tells an extraordinary story about the necessity of creative destruction. This, of course, is the well-known phrase coined by economist Joseph Schumpeter who took from the language of evolution to describe capitalism:

> The opening up of new markets, foreign or domestic, and the organizational development from the craft shop and factory to such concerns as U.S. Steel illustrate the same process of industrial mutation [...] that incessantly revolutionizes the economic structure from within, incessantly destroying the old one, incessantly creating a new one.[10]

At a meta-cognitive level, the level of analysis of biological phenomena, creative destruction can also be identified. One generation's presumptions and conclusions are dismantled by a subsequent generation. This is not to say that what is created out of destruction is necessarily positive, but the cyclical process persists. Throughout this book, I have shown the lines of thought that have changed over the years in various disciplines, depending on social mood; and now we return to that project in assessing certain facets of scientific thought in biology.

## The Evolution of Biology

Before Darwin, and up until the mid-nineteenth century, the high-testosterone biblical perspective ruled the Western academic mindset regarding the origin of life and the creation of species. This benevolent view of existence asserts that an omnipotent, all-knowing, loving, and ethical God created the animal and plant world for the good of his greatest creation, mankind, whom he created in his own image. One of the major biblical themes is the significance of each individual human being as a special creation by God. This is indicated in the high-testosterone, patriarchic culture by a strong sense of self-interest, impelling men to commit in marriage and reproduce their own offspring.

However, testosterone decline during the 1840s and 1850s destroyed that extremely optimistic view of creation and replaced it with a more sinister and malevolent view of nature. The Darwinian battle for existence is antithetical to morality. It is driven by Nietzsche's will to power in an environment of constant war for resources, sex, dominance and submission, conquest and rape, parasitism, and exploitation, and finally the death and mass extinctions of the less fit in the competition for survival and reproduction. This was also the period in which Marxism was developed as theory of class struggle between the oppressors and oppressed in a view of nature as a brutal group contest for dominance status that would yield the most and best resources. Again, the primary change is from the biblical individualistic perspective to collectivistic group conflict between populations in the struggle for existence.

In the history of evolutionary biology, we see periods of negative social mood when the focus of evolution is on a struggle for survival. In periods of positive social mood the interest changes to the study of the information contained by the organisms—the creative code of life, now known as the DNA contained in genes. After a twenty-year period of negative social mood, ending in 1859 and leading to the American Civil War (1861-1865), a wave of rising social mood in the 1860s led to the victory by the North that finally ended slavery. Neil Howe and William Strauss considered the era following the Civil War a first Turning—a High period. During this wave, Gregor Mendel's work on genetics and heredity was published in 1866.

# Sex Wars

Testosterone levels peaked in the late 1860s and gradually began a long decline into the 1930s, causing the rise of socialism and two world wars over the survival of the species. In the early twentieth century, scientists proposed the initial ideas on Symbiogenisis. The prominent Russian biologist Konstantin Mereschkowski first articulated this in 1910 in *The Theory of Two Plasms as the Basis of Symbiogenesis, a New Study or the Origins of Organisms*. He was inspired by research on lichen, an organism composed of algae and cyanobacteria living in symbiosis with fungi. The French scientist Paul Portier published *Les Symbiotes* in 1918, claiming the origin of mitochondria in Symbiogenesis; and in 1922, the American Ivan Wallin proposed similar ideas. The first to apply the terms of Darwinian evolution to Symbiogenesis was Boris Kozo-Polyansky, another Russian scientist, who in 1924 published the book *Symbiogenesis: A New Principle of Evolution*.

In 1942 Julian Huxley published *Evolution: The Modern Synthesis* and originated the term "modern synthesis" as the defining twentieth-century theme in biology. Huxley sought to synthesize Darwin's model of natural selection with Mendel's ideas in genetics in order to establish evolution as biology's primary paradigm. Evolution is described in terms of genetic information that accumulates change throughout generations. One of the primary ideas of this synthesis includes analyzing evolution not it terms of individuals, but populations in which genetic diversity through mutations is the underlying factor to explain change through time.

Testosterone increase during the period of the 1940s and 50s restored an ideology of individualism and optimism, and focus returned again to discovering the genetic code of life. The apex of this research was the famous discovery by James Watson and Francis Crick of DNAs structure in 1953. The major focus of the period 1942-2000 was on the information content in DNA, and one of its famous exponents was Richard Dawkins with his idea of the "selfish gene." The selfish-gene theory posits that the individual is the unit of selection, as a gene that "selfishly" and efficaciously replicated itself during the course of evolution and became successful, whereas genes that failed to work for their own self-benefit failed to compete and were eventually deleted from the gene pool.

A sub-cycle of decline referred to repeatedly in this book came during the 1960s and 70s. In these years, a different interpretation of evolutionary change brought about a paradigm shift. Lynn Margulis published the formative paper "On the Origin of Mitosing Cells" in 1967, proposing her revolutionary thesis of cell merging that became accepted as mainstream science in the 1980s. In 1983 Margulis was elected to the National Academy for Sciences. Also during this period the feminists and socialist scientists developed a new view of nature, called the *Gaia hypothesis* that states the feminine position of Mother Earth. Chemist James Lovelock formed this hypothesis that looks at planet Earth as a living superorganism that regulates its own physiological systems, such as temperature levels, similar to biological complex life. He was joined by Margulis in developing the Gaia hypothesis. This is a comprehensive biological idea with respect to the biosphere, proposing that life forms evolved in ways that actually shape the stability of environmental parameters, such as ocean salinity, global temperature control, levels of oxygen in the atmosphere, and other variables that determine our earthly habitat. Although the hypothesis was initially largely criticized by the mainstream academic establishment, it was later used in emerging fields such as earth system science and systems ecology.

Radical ecologists from the pagan Green movement embraced the notion of Earth as living, Mother goddess. Lovelock headed the global warming fear-mongering campaign as their high priest in his 2006 book *The Revenge of Gaia: Why the Earth is Fighting Back – and How we Can Still Save Humanity,* calling for deindustrialization of the West in order to curb carbon emission and reduce climate change.

Lynn Margulis went on to further develop the theory of Symbiogenesis, suggesting that major transitions of evolution are driven by the same kind of symbiotic interactions that initially transferred genes from bacteria and viruses into a complex eukaryotic cell. Similar interactions build sustainable relationships *between* organisms of different phyla or kingdoms, which lead to the permanent exchange of genetic information between them. While her limited theory in the particular, empirically studied case of mitochondria and chloroplast was accepted, her more elaborate version of a general theory of Symbiogenesis as the origin of complex life remained a fringe idea well into the 1990s. She was deeply critical of the competitive nature of neo-Darwinist theory and its emphasis on gradual accumulation of mutations rather than rapid change through Symbiogenesis. For Margulis, evolution was based more on mutualism and cooperation between different organisms than on intra-species competition.[11] Although the modern synthesis that created neo-Darwinism was conceived in the late 1930s, she predicted in 1990 that neo-Darwinism would prove to be "a minor twentieth-century religious sect within the sprawling religious persuasion of Anglo-Saxon Biology," and further wrote that her fellow biologists would "wallow in their zoological, capitalistic, competitive, cost-benefit interpretation of Darwin—having mistaken him." Neo-Darwinism, she proclaimed, "is in a complete funk."[12]

Moreover, Margulis' disdain for the neo-Darwin synthesis extends to the scientists as well as their methodologies. She minces no words in paragraphs like the following:

> Neo-Darwinism is an attempt to reconcile Mendelian genetics, which says that organisms do not change with time, with Darwinism, which claims they do. It's a rationalization that fuses two somewhat flawed traditions in a mathematical way, and that is the beginning of the end. Neo-Darwinist formality uses an arithmetic and an algebra that is inappropriate for biology. The language of life is not ordinary arithmetic and algebra; the language of life is chemistry. The practicing neo-Darwinists lack relevant knowledge in, for example, microbiology, cell biology, biochemistry, molecular biology, and cytoplasmic genetics. They avoid biochemical cytology and microbial ecology. This is comparable to attempting a critical analysis of

# 5 •THE ORIGINS OF SEX AND DEATH

> Shakespeare's Elizabethan phraseology and idiomatic expression in Chinese, while ignoring the relevance of the English language![13]

Her criticism that mathematic formulas are devoid of empirical basis in biological science, extends to the use of economic language. Cost–benefit terminology "perverted the science with invidious economic analogies."[14] In those words she takes a straight shot at Stephen J. Gould and his "people," who "tend to believe that species only diverge from one another." Whereas she maintains that "species form new composite entities by fusion and merger. Symbiogenesis is an extremely important mechanism of evolution."[15]

## The Holistic Model of Evolution

Due to the testosterone decline since 2000, and particularly after the depression of 2008, the more holistic view of evolution is making further inroads in Western science. In his book *Evolution: A View from the 21st Century* (2011), James A. Shapiro proposes an integrated understanding of evolutionary biology that merges Symbiogenesis, epigenetics, and saltationism (i.e., a significant mutation that pushes evolution forward quickly), recognizing "evolutionary change as active cell process, regulated epigenetically and capable of making rapid large changes by horizontal DNA transfer, inter-specific hybridization, whole genome doubling, or massive genome restructuring."[16]

One example of this shifting model is seen in a 2001 research paper theorizing that cancer was not simply a disease that attacks the body, but an ancient evolutionary adaptation to changing environmental conditions. Aging is programmed into our biology, caused by reduction in the growth hormone testosterone and the rise of stress hormones that cause cells to become cancerous with aging. The problem of partially differentiated cells found in tumors might, according to this hypothesis, be the result of genetic and epigenetic malfunctions. An older layer of genes gets "unlocked," which is none other than the driving code of an ancient evolutionary state of primitive eukaryotes. The idea is that if cancer mutations interrupt and wreak havoc on genetic cooperation, to understand the order by which early multi-cellular code evolved would allow for a kind of reverse engineering of the malfunctions that cause cancer. It would account for "the paradoxical rapidity with which cancer acquires a suite of mutually-supportive complex abilities."[17]

Another example of the more holistic view comes in the transformation of how our bodies are understood, from being single biological systems to an aggregation of multiple cellular agents working in cooperation for our complex bodies and minds to function. The gut–brain axis describes the biochemical signaling network between the gastrointestinal tract and the central nervous system, primarily the brain. Some have expanded it to include the role intestinal bacteria or flora play in influencing the operation of our brains, and also include the endocrine system, the hypothalamic–pituitary–adrenal axis. Interest in this field has exploded, evidenced by the 2014 publication of the study *The Role of Microbiome in Central Nervous System Disorders,* discovering that germ-free mice exhibit an enhanced endocrine response to stress. [18] The gut microbiome is a community of billions of bacteria that live in our intestines and is responsible for multiple bodily functions. More than ninety percent of our serotonin production is made in the gut, and over half of dopamine production, leading to major effects on our brain function through this gut–brain axis. Anxiety and mood disorders as well schizophrenia and autism are associated with deficiency in the gut flora. In addition, the bacteria in our gut have millions of genes, greater by a factor of a hundred, compared with the twenty thousand genes in the DNA of our human eukaryotic cells, so that we live in a symbiotic relationship with our bacteria, complementing each other's needs.[19]

This paradigm shift into a holistic, collectivist worldview that denies individualism is redolent of Eastern mystical philosophies, such as Buddhism, and shows up in more and more bold terms. An article published in the University of Chicago's *Quarterly Review of Biology* in 2012 leads with this title: "A Symbiotic View of Life: We Have

Never Been Individuals."[20] The authors claim that the activity of bacteria and other microorganisms in plants and animals challenges our definitions of what comprises an individual. Such symbionts functioning in metabolism and the immune system are critical to the evolution and sustaining of various creatures—humans included. Therefore, "[A]nimals cannot be considered individuals by anatomical or physiological criteria because a diversity of symbionts are both present and functional,"[21] the authors state. Furthermore:

> Recognizing the "holobiont"—the multicellular eukaryote plus its colonies of persistent symbionts—as a critically important unit of anatomy, development, physiology, immunology, and evolution opens up new investigative avenues and conceptually challenges the ways in which the biological subdisciplines have heretofore characterized living entities.[22]

This conclusion that individual identity is obsolete is hardly imperative. As fascinating and helpful as it is to research the diverse bacterial world within us (or other creatures), their presence by no means necessitates a change in ontological categories. Such research offers us a more nuanced understanding of mechanism, not identity. To say that creatures "cannot be considered individuals" denies how we actually function in the world, which is to say as discreet beings: person, dog, daffodil. Merely because microbes function within and conglomerate dynamics are at work (e.g., epigenetic expressions) does not mean that individual entities cease to exist as such. Rather, what is challenging the definition of "individual" for these scientists is a mindset predisposed to communalism.

The very foundation of Symbiogenesis stands on the socialist worldview of its Russian lineage, and current researchers inherited it from Margulis. This illustrates that differences in political philosophy also stem from the same source as our views in the natural sciences: both are products of how hormones shape our culture. The English, Puritan Christianity that drove the rise to dominance of the Western, individualistic culture tends to view nature as a composite of independent entities, while the socialist worldview is more holistic.

The Russian communist system is a product of the feminine, pagan culture of collective submission to authority. They conceive the origins of sex in Symbiogenesis, whereby the mitochondria are enslaved by the ancient bacteria that engulfed them.

The change in the academic mindset regarding gene-centric individual versus epigenetic and group selection is evident in a 2013 essay by David Dobbs who calls for the end of the selfish-gene perspective of Dawkins et al.[23] He claims that the scientific understanding of genetic dynamics requires a shift to another model of evolutionary development. Because gene expression can be adaptive to swift changes, like the way grasshoppers morph into locusts, or can be flexible, like a genetically identical bee whose gene expression defines her over and against all her sisters as the queen, Dobbs writes, "we need to see the gene less as an architect and more as a member of a collaborative remodeling and maintenance crew." [24] He also offers the analogy that genes are "like managerial teams regulating the behaviour of individual genes for the interest of the organism.[25] It is complexity and variability that Dobbs encourages his readers to comprehend:

> This flexibility of genomic interpretation is a short path to adaptive flexibility. When one game plan written in the book can't provide enough flexibility, fast changes in gene expression — a change in the book's reading — can provide another plan that better matches the prevailing environment.
>
> We need to replace this gene-centric view with one that more heavily emphasises the role of more fluid, environmentally dependent factors such as gene expression and intra-genome complexity...[26]

A more dynamic view incorporates ideas such as gene-culture coevolution and would replace the evolutionary equivalent of the historical "Great Man" model of history. In that nineteenth-century model, individual leaders and innovators like General Dwight D. Eisenhower were credited with the impetus unto social change or historical events. In the case of Eisenhower, for example, he cites the

# 5 ●THE ORIGINS OF SEX AND DEATH

D-Day victory ending WWII. In a revised view, the general's cleverness or leadership savvy did not lead to success so much as the soldiers' abilities to "repeatedly improvise their way out of disastrously fluid situations."[27] Rather than Tolstoy's "masters of history," Dobbs explains, historians began to consider these influential individuals as *servants* of history. Contemporary biologists would likewise revamp the selfish-gene schema to one in which genes are not the motivators of change, they are the "agents that institutionalise change rising from more dispersed and fluid forces."[28] The dynamics comprising such fluid, regulatory networks, it's reported, outnumber genes fifty to one. Genes are awash in them. The natural question, therefore, that Dobbs' article raises is what is actually being selected, individual genes or the "managerial team?" He is clear that it is the team, indicated in the tag line of his essay, "Die, selfish gene, die."[29]

This essay is indicative of the science war between the symbiosis school of mutual cooperation that Margulis and her progeny represent and neo-Darwinists like Dawkins, who conceive of competitive genes that evolve independently, strictly through genetic mutations. Margulis, represents the feminine brain that views cooperation as fundamental for the superorganism to succeed, while Dawkins represents the masculine, competitive brain that views individuals as the basic unit of natural selection. However, evolution works through phase transitions between these two modes of survival. When solar energy levels are high and plant photosynthesis supplies the animal food chain abundantly, organisms are more independent, because they can survive on their own. But when energy levels decline below a certain threshold, a process of cannibalism begins in which one organism seeks to devour another in order to survive; this leads to Symbiogenesis and the creation of new superorganisms.

While there has been a strong tendency in academic thinking to veer toward a worldview that promotes collective, cooperative models, it is most problematic when it gets taken to extremes or considered the sole lens through which to see life. Lovelock's Gaia hypothesis, even in the more limited manner that Margulis accepted it, is typical of the extreme female brain that associates human life with nature, as we see in Eastern mysticism. This extreme position is dangerous for science and philosophy, because it denies objectivity and logic. The problem with cooperation as a driving metaphor is not the observation that individual entities at various levels work in concert to evolve, it is, once again, the anthropomorphizing of a *mechanism*. To state that genes "cooperate" implies that they have a choice not to cooperate. And even when individual genes or epigenetic expression do not cooperate with the status quo—that is, even when we see mutations—the aberrant mechanism had no cognitive agency in the process. Yet when enhanced understanding of processes leads to an all-out rejection of another facet of a process, the rational ability to discuss what is actually transpiring gets weakened. Moreover, the communalist perspective leads to the kind of earth-worshipping, pagan imperialism that degrades mankind. As has already been described in these pages, the likes of Eco-feminist environmentalism literally emasculates the culture, belittling humanity's greatest drives to create and to prosper.

Therefore, in order to understand the full picture of the complexity of nature, we have to integrate both the yin and the yang. What developed throughout the last decades of the twentieth century was a model of multi-level selection. Multi-level selection theory determines that selection occurs both on the individual level, the individual's hierarchy within groups, and group level, the hierarchical structure between groups. In cells the nucleus is composed of gene sequences, or groups of genes, that work together to produce the organism. This means that in reality, each individual is a group of individual entities that fuse together into a single being that is an individual on the higher levels of the hierarchy of life. Because each individual is in essence a group of individuals of the lower level in the hierarchy of group organization, the debate between individual and group selection is fundamentally void of meaning. In complex life there is no selfish-gene, there are only groups of genes that comprise greater groups on a higher level in the ever-increasing hierarchy of organization.

Nature both creates individual entities and organisms and then destroys their independence, subsequently recombining them to create a greater whole that benefits this more complex whole with new characteristics that make it more adapted for survival in its changing environment. This recurring cycle of creative destruction in nature is the origin of the complexity we find in biology. In promoting the cycle view of complex life, Martin Brasier writes that collapse is inevitable as efficiency rises in complex systems. He describes the "boom to bust" phenomenon of ecosystems such as coral reefs that utterly collapse every twenty to thirty million years, as evidenced by the fossils. His vivid language is both informative and amusing, noting that throughout four-hundred-million years "reef ecosystems have flirted dangerously with algae, at first cultivating them like house plants, then like arable farmers, and then like symbioholics."[30] Braisier goes on to draw the parallel between ecosystems that "flip" once they reach a certain peak efficiency and human civilizations:

> Have we humans seen all this before? Well, yes, I think we have: in the collapse of civilizations around the Mediterranean about 1190 BC. And in the collapse of the Classical world in c. 410 AD. And to a lesser extent, in the collapse of 1929. And in the collapse of our modern banking system too.[31]

Braiser's essay exemplifies the model presented in a seminal text that was published in 1995 entitled *The Major Transitions in Evolution*. John Maynard Smith and Eörs Szathmáry set out a straightforward outline of these transitions:

> Smaller entities have often come about together to form larger entities, e.g., Chromosomes, eukaryotes, sex multicellular colonies.
>
> Smaller entities often become differentiated as part of a larger entity, e.g., DNA & protein, organelles, anisogamy, tissues, castes.
>
> The smaller entities are often unable to replicate in the absence of the larger entity, e.g., DNA, chromosomes, Organelles, tissues, castes.
>
> The smaller entities can sometimes disrupt the development of the larger entity, e.g., Meiotic drive

(selfish non-Mendelian genes), parthenogenesis, cancers, coup d'état.

New ways of transmitting information have arisen, .e.g., DNA-protein, cell heredity, epigenesis, universal grammar.[32]

Put another way, a series of evolutionary transitions in individuality (ETI) act as "major landmarks in the diversification of life,"[33] transmitting fitness from the lower levels to new, higher levels: "from nonlife to life, from networks of cooperating genes to the first prokaryotic-like cell, from prokaryotic to eukaryotic cells, from unicellular to multicellular organisms, from asexual to sexual populations, and from solitary to social organism."[34] And it is cooperation that is understood as the fundamental "force" that drives the fluid organizing of life.[35]

From a complex-systems perspective it is difficult to distinguish between living organisms and the emergence of complexity in physical systems. This is even seen in inanimate systems. Voltage applied to a cluster of metal ball bearings within a petri dish causes them to move into intricate structures that look like veins or trees. The online technology magazine *Cosmos* explains that the phenomenon of self-organizing is ubiquitous:

> Self-organising patterns and systems arise everywhere in nature, from the growth of crystals and the folding of proteins to flocking birds and economic markets. In technology, self-organising swarms of robots and drones are a cutting-edge research interest. In both cases the same principles apply: elaborate behaviour can develop from large numbers of simpler elements interacting according to simple rules.[36]

Given this ubiquity of complexity, to comprehend the universe as complex system, we have to understand our minds themselves as complex systems that have both feminine-unifying and masculine-reductive perspectives of the world. Different brain types view reality from different points of view, and that must be taken into account when studying science, which makes it even more complex. Diversity of scientists allows for diversity in perspective. Researchers who investigated the trends in differing scientific biases note that environmental issues are understood differently depending whether rural European-Americans or Native Americans are reporting. The former are more inclined to experience themselves as separate from nature, and the latter as unified with nature. Urban people present another perspective, including the disinclination to even include urban settings in an understanding of ecosystems. Differences in male and female perspectives affect scientific inquiry, as do national and cultural variations. [37] What's critical in the entire enterprise is to recognize that such variations impact perspective, to mitigate bias where possible, but also to appreciate how another way of seeing things can be very informative to how any one observer sees things.

# 6

---

# GENDER, RACE, AND CULTURE WARS

*He has the power to suspend, evade, corrupt or subvert*
*his perception of reality, but not the power to escape the*
*existential and psychological disasters that follow.*[1]
— Ayn Rand

We have discussed the gender and religious wars in previous chapters, but here we reach the even more sensitive topic of race differences and their relationship with hormones, intelligence, culture, and religious conflict. Periods of negative social mood, caused by higher stress levels, often lead to increased polarization as society splits into conflicting groups with opposing ideologies and tribal loyalties that engage in a vicious cycle of war for dominance and submission. One of the major focal points of conflict is the ethnic and racial divide that emerges as individualism declines and societies become alienated on the basis of these collective group identities.

Humanity has evolved over the past tens of thousands of years since Homo sapiens emerged from Africa and spread across the globe to become the dominant species in the world. In that process, many genetic, ethnic, and cultural differences have been accumulating. The Eurasian supercontinent that gave rise to the great civilizations of eastern Asia, such as Chinese and Indian cultures, the Middle East, where the civilization of ancient Mesopotamia, Egypt, and ancient Israel emerged, and the European civilization, from ancient Greece and Rome to Enlightenment Europe.

This great continuous geographical landmass, from England in the west, through the Middle East, to China and Japan in the Far East, has seen the rise and fall of great empires and civilizations, which have intermixed culturally through episodes of war, conquest, and trade, in times of both growth and prosperity, as well as decline into poverty. But, these societies have also had periods of intermixing genetically through interbreeding between different nations and ethnic groups, often through times of conquest and rape. A well-known example of this would be the rise of the Mongol hordes of Genghis Khan in the late Middle Ages, who conquered and subjugated and merged into a single great Empire much of China in the East, the Islamic world in the Middle East, and Russia in Europe. Nevertheless, in spite of some cultural and genetic mixing, there are still conspicuous differences between the East and the West in terms of culture, philosophy, religions, political systems, and even genetic characteristics.

A growing body of scientific research uncovers the East–West cultural gulf in terms of individualism versus collectivism. In this chapter we shall explore the genetic and hormonal differences that give rise to the greater inclination for individualism in the western part of the Eurasian supercontinent in contrast to the east. The West is characterized by a more masculine-type personality than its eastern counterpart because of differences in both hormone levels and hormone receptors in the cells that result in greater influence of testosterone, serotonin, and dopamine. Each culture is also bifurcated from inside into groups with different hormonal types that give rise to a variety of personality traits. In general, however, this more masculine personality type inclines the culture toward greater individualism, independence, and free will, as well as a different philosophical conception of existence. The Western high-testosterone culture tends to more objectivism in philosophy and science, a reality-oriented approach, individual action, defiance and rebellion against authority, and dominance over nature. The feminine East, on the other hand, is inclined to conformity, more social cohesion, obedience to group decrees, submission to natural cycles, mysticism, and belief in a collective fate. This East–West gap explains why Eastern mystical philosophies and religions, such as Buddhism, appeal to many Western feminists, who actually disdain the Western scientific tradition and culture as too masculine and dominant over nature.

Nevertheless, the intensity of distinction between the two extremes is not entirely fixed, but varies according to hormonal cycles that affect humanity as a whole. In periods of global testosterone rise, the West is more sensitive to rising levels of growth hormones and will experience rising growth levels of individualism than the East, as we have seen since the Enlightenment in the seventeenth century. But this is a double-edged sword, as the cycles work both ways. When global testosterone levels decline, the West, having risen to the top of the global hierarchy, will be more susceptible to its complex, individualist social and economic structure collapsing. The transition from the modern, individualist society back to a more collectivist, primitive one may be much more difficult for those who have attained and enjoyed the up cycle and will lack the cultural cohesion to relieve the pain in the down cycle.

If we view Western civilization as a more masculine society than the East, this could explain why the West can collapse in periods of testosterone decline, while Eastern cultures such as China can replace their ascendency. The Eastern Roman Empire in Byzantium, for example, outlasted the fall of Rome by nearly a millennium. There could be a parallel here as to why women live longer than men, as men are more disposable, given they are more vulnerable to hormonal growth cycles. Research shows that not only do male humans but also male animals live for shorter spans than females or even neutered males.[2] Growth and reproduction would have been a genetic priority for males. Professor Thomas Kirkwood of the University of Newcastle in England states that "Our genes treated the body as a short-term vehicle, to be maintained well enough to grow and reproduce, but not worth a greater investment in durability when the chance of dying an accidental death was so great." The chances of "reproductive success" are higher if female bodies are more durable and less susceptible to disease and decay.[3]

From an economic perspective, in the event of collapse, China, which had never developed a fully capitalist system like the United States, would not have to contend with the same scale of transition back to a collectivist system. We see that testosterone decline is causing much of the West to lose what sociologist Max Weber identified as the social qualities of strong family values, long-term planning, and high savings rates along with investment for future growth. The increasingly feminized culture seeks consumption over production and spending over saving. Such behavior aligns with Keynesian economics that calls for big government spending to "save" economic growth as "folk medicine" against recession. In the past forty years, Britain and the United States have been losing their industrial bases. The great automobile manufacturing capital of Detroit is now largely bankrupt and neglected, controlled by socialist labor unions that produce more red tape and administrative wasteful regulations than cars. Mitigating the decline of the manufacturing sector, the one bright spot that can be seen is the innovative drive of the high-testosterone Western culture in high technology in the United States that has created enormous wealth and productivity.

Yet, as manufacturing has declined, New York and London have become the global capitals of finance. These global bubble-creation machines increase artificial debt and induce spending and consumption in the West. They are headed by feminist, socialist central bankers who seek to lower interest rates continually in order to accommodate the instant gratification culture of endless consumption. The inevitable day of reckoning is merely being postponed, but it will come: the explosion of the ever-growing debt bubble. When interest rates finally rise, the West, lacking the productive capacity to serve its huge debt or the moral and cultural backbone to sustain social strife, will self-destruct economically and socially.

Economists studying patterns in socioeconomic trends have utilized the latest research in genetics and neuroscience to construct economic growth models. The have recognized that individualism–collectivism is a significant lens through which to view a culture's economic propensities, specifically "that more individualism leads to more innovation because of the social rewards associated with innovation in an individualist culture."[4] While collectivism promotes efficiency in productivity, economists from Berkley note that individualism has a "dynamic effect" on growth.[5] To be able to assess predictors of economic productivity with a greater understanding of genetic influences is a powerful tool. To see some of the particular genetic influences that drive cultural engines, we look again to hormones.

### Serotonin

Researchers in behavioral genetics have tested participants in various countries for potential differences in neurotransmitting genes. Different forms of a gene are called alleles. The serotonin transporter gene has the form of either an L allele or an S allele. The latter makes its carriers more prone to depression in the face of environmental stressors like interpersonal threat, conflict, or loss. The S allele "is associated with increased negative emotion, including heightened anxiety, harm avoidance, fear conditioning, [and] attentional bias to negative information."[6] The differential between East Asian and European carriers was determined in a meta-analysis that considered over fifty-thousand individuals across twenty-nine countries. The results showed that seventy to eighty percent of East Asians have the S allele of the serotonin transporter gene, while forty to forty-five percent of Europeans carry the S allele.[7]

East Asian cultures function as collectivist much more than Western societies, differences that have been evident throughout much of recorded human history. This can be seen in ancient literature if one compares the fifth-century BCE writings of the Torah and Socrates to the Chinese writer Lao Tzu, founder of Taoism, for example. Interconnection, inclusion, and defined social order all factor significantly in the Asian mindset, creating a beneficial society for those who carry the S allele. A literature review out of the UCLA psychology department in 2010 summarizes:

> Consistent with this notion, there was a correlation between the relative proportion of these alleles and lifetime prevalence of major depression across nations. The relationship between allele frequency and depression was partially mediated by individualism–collectivism, suggesting that *reduced levels of depression in populations with a high proportion of social sensitivity alleles is due to greater collectivism.* These results indicate that genetic variation may interact with ecological and social factors to influence psychocultural differences.[8] [emphasis added]

In considering why such distinct genetic differences evolved with these cultural variables, scientists posit that it could have been caused by disease prevalence. Increased pathogens in East Asia for diseases like typhus and malaria may have led to "increased collectivistic values" by selecting for the S allele.[9] When carriers of the S genotype are supported in appropriate environments, sensitivity to such pathogens is minimized, and not least by the decreased propensity for anxiety and mood disorders that show up at higher rates in Western nations with only half the number of S carriers.[10]

The adaptation of a serotonin transporter gene that increases sensitivity to societal conflict in its various forms and thereby inclines people to a more cooperative societal model is a strong indication of gene–culture coevolution. The genotype, be it S or L, is considered a "singular predictor of cultural values of individualism–collectivism across nations....and is fundamental to any comprehensive understanding of culture."[11] More specifically:

> Recent behavioral evidence indicates that individuals carrying the S allele exhibit stronger attentional bias for negative words and pictures, whereas individuals carrying the L allele demonstrate a stronger attentional bias towards positive pictures and away from negative pictures. By extension, S allele carriers may be more likely

to demonstrate negative cognitive biases, such as engage in narrow thinking and cognitive focus, which facilitate maintenance to collectivistic cultural norms of social conformity and interdependence, whereas L allele carriers may exhibit positive cognitive biases, such as open, creative thinking and greater willingness to take risks, which promote individualistic cultural norms of self-expression and autonomy.[12]

## Dopamine

In addition to serotonin transporter genes, dopamine receptor genes correlate along the independent–dependent or individualism–collectivism continuum as well. The findings are similar. In one study of 398 participants, carriers of the dopamine receptor gene DRD4 in its 7-repeat and 2-repeat alleles showed high levels of cultural-norm adhesion.[13] It has also been found that positive selection of such culture-related genetic changes has increased exponentially over time with population growth and with animal and plant domestication. The more environments changed, the more specified genetic alterations evolved.[14] The example of such gene–culture coevolution presented in chapter 3 is instructive again here: As domesticated herding and milking increased in various societies, genetic adaptations allowed for better digestion of milk, thereby furthering the development of activities supporting human consumption of animal milk for nutrition.[15]

Dopamine is the motivating hormone and plays a large role in how the brain learns, reinforcing with positive feelings that create a behavior feedback loop. The process called "reinforcement-mediated social learning" comprises "a set of mechanisms that enable the organism to select behavioral options that maximize anticipated rewards. These mechanisms include discerning of behavioral patterns, selection of one's behaviors, and tracking of the reinforcements given to these behaviors."[16] The social rewards that stimulate dopamine, paired with a genetic makeup that increases dopamine sensitivity leads to increased "cultural acquisition."[17] This was exemplified in the study referenced above, in which carriers of DRD4 in the 7-repeat or 2-repeat alleles are found to have increased dopamine signaling and have been found to align more strongly with their own culture's tendencies on the independent–interdependent spectrum. Noncarriers of these alleles do not indicate strong social orientation one way or the other.[18]

Correlating with independence–interdependence is evidence that the degree of dopamine reception could affect people's political bent. A study of almost 1800 Han Chinese female students in Singapore revealed that the participants with two copies of the DRD4 4-repeat allele reported as "significantly more conservative" in their self-reported political views.[19]

## Testosterone

Given that serotonin and dopamine sensitivity align with distinctions in cultural mindsets, it is no surprise to see that testosterone reception does as well. The endocrine system is highly integrated in ways we continue to learn about. As recently as 2015, testosterone was first shown to directly affect the number of serotonin transporters in human brains. Previously in this book the statistic was cited that men have a fifty-two percent higher rate of serotonin uptake, which is a clear indication of the relationship of testosterone levels to serotonin receptivity. As a result of this connection, testosterone has also been found to enhance treatment for depression in conjunction with SSRI medications. Increased serotonin transporter levels can be measured after only one month of combined therapy with antidepressants and testosterone because of the greater number of potential serotonin binding sites. In fact, the serotonin levels are "significantly higher" in that time frame and improve even more with continued treatment.[20] Other studies have shown that certain biochemical precursors to testosterone levels are positively correlated to depression and anxiety. One study looked at men fifty-years old and older and confirmed that panic and phobic anxiety levels are higher in men with fewer androgen (i.e., male sex hormone) receptor cells.[21]

For testosterone to carry out such functions as improving serotonin reception, its own receptors must be ample and functional. Without adequate testosterone receptors, for example, a person with the XY chromosome combination—a male—will develop as a female.[22] Less extreme examples of testosterone sensitivity exhibit themselves rather definitively across ethnic groups, resulting in differences between body muscle mass and male testes sizes, among other things. An enzyme called 5a-reductase converts testosterone into a more compatible form (DHT) for the androgen receptors. Another aspect of the process pertains to what is termed *CAG repeats* in cells. Those people with fewer CAG repeats produce more of the amino acid glutamine, which in turn increases the efficacy of androgen receptors, hence capturing more testosterone.[23] Years of research has reiterated at a genetic level what can be seen with the eyes: People of African descent, and especially men, have the greatest 5a-reductase concentration and the shortest CAG repeat lengths. Thus, they show the most testosterone sensitivity, manifesting in the highest ratio of fat-free body mass to fat mass, height, and skeletal structure. European, American and Semitic Caucasians (i.e., Jews, Arabs, and Persians) have a lower level of testosterone sensitivity, and Asians the lowest.[24][25]

Although the evolutionary purpose is not thoroughly understood, another feature that correlates with race and testosterone levels is testes size. Researchers weighing Chinese and Danish testes during autopsies found that the Danish men's testes weights was more than twice that of the Chinese.[26] The size of the man's body in this and other studies does not seem to correlate to the testes weights necessarily; but the race of the man is clearly an indicator. In fact, the difference in testes weight is "far beyond what average racial differences in body size would predict," with those weights averaging as follows: 24 grams for Asians, 29 to 33 grams for Caucasians, and 50 grams for Africans.[27] Greater testes weight means more sperm production, which raises the question why the races have evolved with these differentials. A primary theory focuses on the variations in mating behavior between cultures. In the animal kingdom, creatures with multiple mates (i.e., polygamous) have larger testes with respect to their body size, creating more sperm that is more active than monogamous creatures.[28] This follows when we consider that more effective sperm—in amount and quality—is more likely to be successful in mating. Oxford zoologists Paul Harvey and Robert May state that the less monogamous a species is, the greater need for sperm. The cultural adaptations for this can be imagined by looking at the Danish and Chinese example. According to Professor May, "It could be that for generations the Chinese lived in a society that was sexually secure, while over the same time the Danes lived in a rape-and-pillage and violence kind of society."[29]

While at face value this would seem a logical enough component of an answer, this hypothesis is considered controversial, as it does not account for a range of contributing factors:

> [S]ome of the reluctance to connect culturally sanctioned behavior with genital anatomy is as much due to the difficulty of finding reliable historical information about true rates of female promiscuity as to the emotionally charged nature of the material itself. Furthermore, diet and environmental factors would have to be factored in before arriving at any solid conclusion regarding the relationship between sexual monogamy and genital anatomy. For example, many Asian diets include large quantities of soy products, while many Western people consume large quantities of beef...[30]

Furthermore, the scientific and political environment for decades has made people skittish when it comes to any kind of cultural assertions based on biology, because there is always someone who takes the information to extremes. In the case of testes size and race, if it is claimed to have evolved in direct relation to promiscuity, and African men have the largest testes, then this can be used to draw a disparaging picture of African people or those with African heritage. The biological facts lead some to moral conclusions that are less than flattering. An infamous writer who did just that was the professor J. Philippe Rushton of the University of Western Ontario. Rushton was prolific in his pronouncements that genetic and physiological differences between the races constitute a hierarchy "in which blacks rank as the least intelligent, least altruistic, least nurturant as parents, most licentious, most criminal and most given to multiple births ('litters,' in his terms)."[31] Rushton and those who concur with his conclusions consider African-heritage people "the most animalistic" among human races and Asians the "most highly evolved."[32] Given that there is such a tendency to negatively characterize on the basis of biological distinction, no area is more controversial than that of race and intelligence, which we will address next.

## Race and Intelligence

Measurements of intelligence is perhaps the most problematic topic in the social sciences, largely because of its popularity among actual racists, fascists, and National-Socialists who call for race wars and even genocide. However, a primary theme in this book is that the current unscientific method of disregarding our biology in order to promote our ideals by disconnecting them from reality will not save civilized society. As a paraphrase of Ayn Rand goes, "We can ignore reality, but we cannot ignore the consequences of ignoring reality."[33] To transcend our biology we have to courageously face and study the facts in order to learn to control our own future, much of which is coded in our genes and enacted through our hormones.

Hence, we will ignore the dictums of political correctness and study this subject objectively. As I have already established, there are differences between individuals in every society and culture that do not cause animosity in periods of positive social mood, regulated by rising levels of happy chemicals in our brain. We should be more concerned with the fall of testosterone and the negative trend in social mood that drives us into social division, group conflict, and war than with our genetic differences, because hormones epigenetically activate our inherited tendencies to form collective groups that fight for the available resources in times of crisis and scarcity.

There are acute differences between individuals within groups, as well between groups. Both the pressure in individual selection within groups and group selection in intergroup conflict only rise in periods when our brain reacts to signals of stress, signaling us that we should fight for our dominance and survival in difficult times of lack in resources. Therefore, the cure for fascism and race wars is not by ignoring nature but through conscious effort to transcend its evils and rationally choose the good. In times of social harmony and a culture of growth, economic growth is actually driven by differentiation and specialization, when individuals can work together to form complex social networks to the benefit of the whole society as well as its parts. Thus, our genetic and cultural difference is to be celebrated as the factor that enables diversity and opportunity to fill new niches.

The differences in IQ tests between races in worldwide studies average as follows: People from sub-Saharan Africa have an average score of 70; black-Americans have a score of 85 (due to interbreeding with whites); Caucasians, 100; and East Asians 106.[34] East Asians including the Chinese, Japanese, and Koreans, have the highest IQs and the lowest testosterone levels of these three groups. While the Europeans score on average a bit lower, they have higher testosterone levels than the East Asians.[35] There is one notable exception to these scores, which is that Ashkenazi Jews, those Jews of Eastern European heritage, versus Sephardic or Oriental Jews, have the highest overall average at 110, with particularly high verbal intelligence at a score or 125. This will be addressed further later in the chapter.

Given how these intelligence quotients present, it is ironic that some racist white scientists who believe in white separatism—ethnic separation of Europeans or European Americans from others groups on the basis of race—actually admire the ethnocentrism exhibited by the East Asian cultures. In so far as those cultures tend to reject foreigners and embrace collectivist ideologies, some white racists actually view the East Asians as their model for a so-called "super race."

# 6 • GENDER, RACE, AND CULTURE WARS

In theorizing about how people groups developed differently in the realms of brain size, body sizes, and varying cognitive abilities, scientists have postulated that migration out of the African savanna could have led to selection for larger brains to accommodate increased pressures of survival in colder climates. More complex shelters and clothing were necessary; challenges with food preparation and storage for winter, and the transition from a gathering, nomadic culture to a stationary agrarian one might all have been factors in the eventual European and Asian peoples evolving for higher cognition such as IQ tests assess.[36] Further, the prey that would have endangered northern cultures as well as the animals hunted were smaller and slower. Instead of elephants and lions, cheetahs and gazelles that tend to move in larger groups across wide-open spaces, Europeans and Asians contented with bear, wild boar, and wolves, more likely in forested territory. This would have diminished the need for as high a muscle-mass ratio as the Africans had and other characteristics of higher testosterone, such as higher aggressiveness and impulsivity—all strong features of nomadic, hunting, warring people groups.[37]

Testosterone drives variability in intelligence regarding gender, as well. Many studies have found that the degree of IQ variance is greater among males than among females. Female intelligence is more equally distributed, whereas in males, there are more geniuses on the high end and more men of low mental aptitude on the other end. This is another manifestation of testosterone's property to produce inequality and drive variability: men compete to find females to mate with, while nature takes fewer chances with women, who tend to be less diverse as a group.[38] This is also true regarding differences in variability between ethnic groups. East Asians have been found to have lower testosterone, and hence, like women, less variability in the population distribution of IQ scores than Caucasians. Therefore, although East Asians are on average more intelligent, the West produces more geniuses at the extreme ends of the distribution curves. For instance, at the near-genius level (an IQ of 145), brilliant men outnumber brilliant women by eight to one.[39]

Furthermore, while East Asians display higher intelligence, their culture has not produced the level of ingenuity and innovation of Europeans since the Enlightenment. This is likely attributable to the higher testosterone levels in Europe that led the West to embrace individualism and capitalism and a culture that promotes objective scientific discovery. Moreover, the intellectual challenges in framing novel philosophical worldviews and innovations in science, art, and technology require the personality characteristics of independence and even a kind of rebelliousness. Therefore, while social conformity is positive for social cohesion in the East, a culture higher in testosterone than the East Asians and higher in IQ than the Africans created the optimal environment to create new ideas, methods, and scientific discoveries that continue to challenge the old, entrenched, accepted social norms and beliefs. This was the capability that moved Western society from an agrarian to an industrial and then technological culture and economy.

## Racism and Fascism

The white racists in the West, many of whom claim to be atheists, actually maintain a pagan-fascist worldview that is similar to Nazi eugenic theories of race wars. This view sees the world as a racial battle for the survival of the superior white master race against the Jews, blacks, and other "inferior" ethnic groups. Ironically, given that some Jews have equal if not higher IQs than the Caucasian race, this would indicate that the white racists do not hate others groups merely out of disdain for those with a lower intelligence quotient.

The particularly high verbal intelligence in Ashkenazi Jews is notably in contrast to their deficiency in visio-spacial skills, a possible genetic tradeoff in the allocation of brain power. This explains why Ashkenazi Jews do extremely well in academics, mathematics, and other fields that require increased problem solving and abstract reasoning, but are less represented in manufacturing industries that require more visio-spacial skills. One theory proposes that Jews in Medieval Europe maintained a high degree of cultural cohesiveness and inbreeding that led to some genetic diseases, but also enhanced their intelligence to excel the skills they valued.[40] Furthermore, the Talmudic educational tradition that trained the minds of Jewish boys from a young age and has been practiced consistently for centuries would have perpetuated selection for verbal intelligence in reading, writing, and reasoning skills.

Kevin MacDonald, a retired psychology professor from the University of California Long Beach, takes these ideas about the environment that fostered selection for high-verbal IQs among Ashkenazi Jews and posits a pernicious twist that repeats a classic anti-Semitic slander: the Jews evolved as a parasitic group, involved mainly in commerce and banking, to prey on the productivity of the European people, their host culture.[41] This was the tenor of disparagement famously captured in Shakespeare's *Merchant of Venice* character Shylock as well as the classic slander embraced by the Nazis, among others throughout history. In his book *A People That Shall Dwell Alone: Judaism as a Group Evolutionary Strategy* (1994), MacDonald describes what should be understood as strengths in the Jewish culture—strong group cohesion supporting "practices aimed at producing high intelligence and high investment parenting ideally suited to developing a specialized ecological role within human societies"[42]—and characterizes them as "eugenic" and "cultural manipulation of segregative mechanisms."[43]

MacDonald developed his evolutionary theory about the Jews-as-parasites out of his studies of wolf packs. His views on human group conflict and group identity politics are based on animal models and are a result of low-testosterone that has led to the spread of a collectivist worldview. His fascist worldview, accompanied by racism and white supremacy, conceives of the world as a story of group conflict rather than the prospect of peaceful coexistence between people of differing cultures and ideas. In attributing the high intelligence and ingenuity of Judaism to a "group evolutionary strategy," he perpetuates the blame and scapegoating of Jews that is especially prevalent in periods of testosterone collapse. Such times led to the implosion of financial and economic bubbles, as well as increased social strife and group conflict in times like the first decade of the twenty-first century and during the Great Depression of the 1930s.

Furthermore, in eras when the Jewish patriarchy loses its dominance, the people lose their group cohesion, and there is a sharp rise in interfaith marriage. This trend repeats through history. In the period of Weimer Germany (1919-1933), there was a forty percent rise in intermarriage for Jews. Once the Nazis took power, the practice of marrying outside the faith all but ceased.[44] In a 2013 poll, the intermarriage rate for all US Jews was fifty-eight percent and seventy-one percent for non-Orthodox Jews. This can be compared to a seventeen percent mixed-marriage rate in the US before 1970. All told, this is a function of the general decline of Jewish adherence to the historic faith in the US, with only one-third belonging to a synagogue and twenty-five percent of Jews reporting they do not believe in God.[45] Moreover, these declines in group cohesion for Jews correlate with both the first and second waves of feminism in the 1920s and 1970s respectively. Lacking male leadership, women become less interested in males of their ethnic or religious group. This trend is particularly noticeable in the decline of Jewish faith, because it is based on the masculine ideal of God the Father and family values, which are attacked by the feminist-communist pagan culture that is anti-God and anti-masculine.

There are multiple ways to refute perspectives like MacDonald's, and one interesting contrast to his kind of ideas is to look at a group of Jewish people who came to the British colonies in America during the seventeenth and eighteenth centuries. Most of these colonial Jews were Sephardic (though many were Ashkenazi as well) and had developed the same strong cultural cohesion and skills in commerce that the Ashkenazi are known for and, in MacDonald's case, are cynically attributed to some grand genetic plot to dominate. Situated in the environment rich with ideas of liberty for all mankind and tolerance for religious differences, the Jews in colonial America were remarkably prosperous economically and socially. Compared to Europe in those same centuries, the degree of anti-Semitic bias, while not lacking, was distinctly less; and many Jews in the late 1600s all the way through the American Revolution and onward distinguished themselves as respected farmers, merchants, doctors, soldiers, and statesmen. The atmosphere of the Enlightenment in colonial America allowed for thriving synagogue communities from Rhode Island to Georgia with congregants who encountered far less of the hatred, violence, and shunning experienced by their European counterparts. The case could be made that it was the *ideals* pervasive in the New World that diminished intergroup conflict and allowed for Jews to be an important part of the larger entity developing, which was the United States of America—not as parasitic leaches on a culture, but as joint builders of a culture in which their intelligence and industriousness was respected and fostered.

**Honor Culture**

The flourishing and prosperous economic environment in the West since the Enlightenment has certain problematic downsides. In chapter 2 we cited the gradual loss of stress response in societies with high standards of living and law and order. Immediate threats to safety and sustenance provision are minimized and violent, aggressive behavior lessens. In the case of the West, this has brought about the weakness of the feminized Western culture today in the face of the looming dangers of the barbarian Islamic invasion. The reduced levels of stress and aggression result in low levels of cortisol and testosterone, particularly in academia, which is a very protected environment, where employees enjoy great resources and freedom. However, this leads to a withdrawal of basic masculine instincts, like the ability to protect territory and to defend against foreign threats.

The culture that most promotes masculine virtues in the United States, for example, exists in more rural areas and the South. Various studies show that men from this region exhibit greater tendencies for "an interdependent feedback mechanism between testosterone and aggression that is modified by experiences of victory and defeat...."[46] These men are more inclined to respond to insults out of a strong sense of honor and defense of their ideals than northerners or urban dwellers.[47] One set of studies with students at the University of Michigan looked at eighty-three white young men raised in either the South or the North and how they reacted to being bumped into and insulted.[48] Their responses were assessed by both observed and self-reporting reaction as well as measurement of their cortisol and testosterone levels concurrent with and following the incident. The reaction assessment was how relatively "amused" or "angered" the man was by the situation. Northerners were "relatively unaffected," while those who grew up in the southern "honor culture" responded with more aggression.[49] The Southerners reported that their reputations were threatened, and by a 2.5:1 ratio were more angered than amused. The Northern students, on the other hand, reported amusement more than anger at a 4:1 ratio.[50]

In another study, researchers submitted the theory that such concern for reputation, if it were more acute in the South, would show up in the actions of Southern US presidents more than those of Northern presidents. Their investigation compared President John F. Kennedy to his immediate predecessor Lyndon B. Johnson and matched against other presidents and revealed:

> Interstate conflicts under Southern presidents are shown to be twice as likely to involve uses of force, last on average twice as long, and are three times more likely to end in victory for the United States than disputes under non-Southern presidents.[51]

Many other studies have confirmed these kinds of results. The fact that such cultural distinctions can be measured is one thing, but how they are interpreted or how they breed subsequent theories is another. Generally, in the academic research climate of the past few decades, the qualities of honor cultures, especially how men are trained, are viewed negatively. The men are characterized as insecure, defensive, and "emotionally unstable."[52] One theory by its very name depicts the bias of researchers: low-status compensation. The presumption is that low social status impels people to "compensate" with increased aggression and violence. They impose a kind of

Napoleon complex on whole swaths of society to account for higher rates of violence. A study sampled men in 93 countries and many counties in the South to see if *herding* cultures were more likely to adhere to honor codes that, in turn, promoted more violence.[53] The connection was verified, but the reason given was not that men were responding in accordance with their honor codes, rather that they act out retributively to compensate for the stigma of low social status. In this theory, everything apparently comes down to a wounded ego and low self-esteem.

As we have seen in other areas, the elite establishment that is tasked with most analysis of a culture is increasingly biased against traditional masculine culture where boys are trained to have pride and stand up for their values and to defend their families and communities. When parts of a culture are less inclined to succumb to the systemic process of emasculation, people are effectively ridiculed. These were the people who, in the 2016 US election, Hillary Clinton labeled as "deplorables." Arguably and ironically, that insult cost her the presidential election. The collectivist hive mind in Western academia has developed a monolithic understanding of what motivates men to defend their ideals, reputations, and honor. Their egalitarian ethos means that all aggression is equal. There's no distinction between fighting for freedom and fighting to oppress. When all violence derives from low self-esteem, the criminal or the terrorist becomes the victim of a bigoted or economically oppressed society. This is a meme heard in the West repeatedly since 9/11, for example, by the left: that the US and its allies created the global environment that fosters Islamic retaliation:

> As hungry replicators eager to remold the world, ideas often turn their ultimate weapon—the superorganism— into a killing machine. And, contrary to the doctrines of some modern critics, they do not engage in this "hegemonic imperialism" only in the purportedly "malevolent West."[54]

These are undoubtedly complex issues, but it's valuable to review them from another perspective. The social commentator Harold

Bloom, sets Islamic violence in its historical context that takes into account its foundations and almost 1,400 years of entrenched ideology. Citing the anthropologist Ruth Benedict, Bloom illuminates her notion that the founding personality of a religion or culture exerts enormous influence on how that culture develops.[55] This is a platonic ideal, Bloom reminds us, whereby "Plato is outright convinced that the personality of the leader stamps itself on the society he leads."[56] Islam "imprisons itself" in the pattern of the warring Prophet Mohammed:

> In 629 AD, Mohammed and his warriors attacked and defeated the Jews of three tribes: the Banu Qaynuqa, the Banu Nadir and the Banu Quraiza. Then Mohammed inflicted war crimes on these three Jewish tribes. For two tribes, he commanded ethnic expulsion. In addition, he stole these tribes' wealth and property. In the case of the Banu Quraiza, he committed genocide. He had every male of the tribe old enough to have pubic hair beheaded in front of him in the market place of his headquarters in Medina. Then he distributed the women of the Banu Quraiza as sex slaves, being careful to take the most beautiful and highest ranking as a bride for himself.
>
> In the long run, the forces of Islam conquered or seduced the biggest empire in world history, an empire 11 times the size of the conquests of Alexander the Great, five times the size of the Roman Empire, and seven times the size of the United States. Islam continues to hold most of these lands in its grip today.[57]

This "ravenous superorganism" is perpetuated by what we examined in chapter 2 as a powerful form of impressing submission on Muslims—their call to prayer. Bloom calls this "Mohammed's cleverest meme-hooks, his most innovative conformity enforcers,"[58] and by such continual indoctrination, Muslim supremacy in the world pecking order was held for some 1,200 years, into the early 1800s. In his lifetime alone, Mohammed led twenty-seven military campaigns of the sixty-five he commanded; he "idealized killing unbelievers, and preached the violent takeover of the entire earth."[59]

Power vacuums in geopolitics, such as now exist in the China–US economic "megapower" struggle, allow for systemic openings for Islam to revive its supremacy; and its strategy is the global jihad perpetuating terror attacks from Boston to Bali. Bloom also asserts another angle on this megapower struggle. China, he states, is biding its time until the US and Islam "bleed each other into exhaustion, allowing the growing Chinese mega-power to step in and become the new hegemon, the new alpha superorganism in the global pecking order..."[60]

It cannot be emphasized strongly enough that the mindset of Islamic jihadists or that of a communist super-army at the ready is not the same as the ideals of individual liberty that led to the major conflicts on North American soil. One, to secure liberty from the crown, and the other to secure liberty from the oppression of slavery. The founders' ideals of life, liberty, and the pursuit of happiness came from a high-testosterone, individually expansive motive as opposed to the herd mentality of primitivism that propels communism or jihad. Yet, even in light of these high ideals, there is no such thing as a utopia. As my work in three books has elucidated, social moods cycle up and down throughout history. Sociologists and evolutionary psychologists like Steven Pinker want to present a utopia wherein progressivism in the industrial–technological world has moved humanity beyond violence. However, political philosopher John Gray concurs with the Testosterone Hypothesis:

> Improvements in civilization are real enough, but they come and go. While knowledge and invention may grow cumulatively and at an accelerating rate, advances in ethics and politics are erratic, discontinuous and easily lost. Amid the general drift, cycles can be discerned: peace and freedom alternate with war and tyranny, eras of increasing wealth with periods of economic collapse. Instead of becoming ever stronger and more widely spread, civilization remains inherently fragile and

regularly succumbs to barbarism. This view, which was taken for granted until sometime in the mid-18th century, is so threatening to modern hopes that it is now practically incomprehensible.[61]

This returns us to the question of what it is that drives these waves of social change and its attendant question, raised in chapter 1, how and why do we organize into groups? What propels the superorganism of China to steer its economic organization toward capitalism, for example, or the culture of the West to career down the hill headlong into socialism? In the debate about individual versus group selection, Howard Bloom conceives a kind of amalgam. In his 1995 book, *The Lucifer Principle*, he sought to disclose that humans are "both selfish competitors and pawns of the social group."[62] The group follows patterns of dominance hierarchies *because* of how individuals have selected genetically. Individuals are programmed to integrate with their immediate culture by genetic characteristics that Bloom calls "conformity enforcers."[63] More nuanced than the mechanisms that keep ants and bees conformed to their roles, conformity enforcers in humans instill limitations and alterations to our very perceptions. They assure "a collective perception, a socially constructed view of reality which influences both childhood brain development and adult sensory processing, and which produces a *Weltanschauung* displaying many of the characteristics of a shared hallucination."[64]

The complex dance of hormonal interacting within each person—from infancy—attracts and respells one person to and from another: infants and parents bond, brain tissue adapts to one set of language sounds over others, positive feedback loops train a child to connect to his or her own people and to recognize strangers. Forces both internal and external compel us to conform. There is an "urge to belong" paired with a "fear of social ostracism."[65] In his essay *Beyond the Super Computer: Social Groups as Self-Invention Machines*, Bloom expounds further on the "comparator mechanisms" that communicate our degree of usefulness to our respective communities.[66] Ultimately, a cascade of signals lead to aging and death, including socio-cognitive signals that even repel others from the diminishing individual. He explains that an over-abundance of glucocorticoids and endogenous opiates serve as "deactivators" that impair memory and suppress the immune system. Yet those people still active and enlivened by testosterone and serotonin that maintain their health and energy, their sex drive and their "independence of perception," actually are experienced by others as "socially captivating."[67] Physical and verbal cues resonate with certain individuals who naturally become the leaders and opinion formers of the group. Conformity enforcers and comparator mechanisms work together to give groups "a shared perception of the world, a view so distinctive that it can give outsiders the impression of a mass delusion."[68]

### The New Global Dominance Hierarchy

Readers of this book will recognize in Bloom's model ideas from earlier, such as Jonathan Haidt's "shared intentionality" from chapter 1 or the Male Warrior Hypothesis from chapter 2—all contributions to our endeavors to understand human evolution unto variations of the hive mind or swarm intelligence. Integrating the Testosterone Hypothesis with the idea of the global brain explains why feminists seek to subdue and replace Western men with Islamic jihadism. Testosterone decline signals the aging of the West and the Western male, who has been the global alpha male, having a dominant position in the world since the rise of the European Enlightenment. As in aging apes that are challenged by the new alpha male on the block, now the feminists are joining the barbarians in their holy war to destroy the Western male by conquering the civilization he built.

The feminist multicultural agenda drives females in the West to reject the notion of Western superiority, in the context that all cultures are similar and no culture is better than another. After all, by the secular postmodern ideology, there is no objective ethical measure of good or evil. Hence, Western males can easily be substituted with others ethnicities and races, and the genes of humanity can be mixed through intermarriage and interbreeding. This has been the case in the past, as when Homo sapiens emerged from Africa tens of thousands of years ago and interbred with the Neanderthal population that had been living in Europe for four hundred thousand years.

Only until recently, when testosterone levels were still high, Homo sapiens were considered a distinct and superior race. This informed the isolationist approach of Western dominance. The theory was that the race originated in Africa and spread across the globe, causing the mass extinction of the Neanderthals and other inferior and more primitive hominids, but not interbreeding with them. Now that our dominance and status hormones, testosterone and serotonin, have declined dramatically, we no longer feel superior to other hominids and are willing to admit that interbreeding actually had much beneficial impact on the development of our modern species. New methods of testing ancient remains indicate that up to twenty percent of our genes came from Neanderthals.[69] DNA studies indicate that our resistance against certain illnesses might have come from Neanderthal heritage.[70] Such ancient interbreeding is actually at the root of multicultural ideology, because if Homo sapiens is not a "pure" race, then neither is there such a thing as a pure Caucasian or pure Asian. In the following chapters, we will broaden the lens and look at forces behind our evolutionary development from a literally "cosmic" perspective.

# 7

---◗◦◖---

# HOW SOLAR CYCLES DRIVE OUR EVOLUTION AND HISTORY

*The thing that hath been, it is that which shall be; and that which is done is that which shall be done: and there is no new thing under the sun.*
— Ecclesiastes 1:9
*By the Law of Periodical Repetition, everything which has happened once must happen again and again and again-and not capriciously, but at regular periods, and each thing in its own period, not another's, and each obeying its own law... the same Nature which delights in periodical repetition in the skies is the Nature which orders the affairs of the earth.[1]*
— Mark Twain

To this point in the book, the presentation has been about trends in evolution and social mood, but the undergirding story of all these activities is more fundamental and more profound. It is the story of our planet's relationship with its star, the sun. All life on earth depends on free energy from the sun to fuel its growth and reproduction. However, this constant and reliable flow of solar radiation is far from secure, as science has discovered by studying the past history of our planet. Periods of global cooling—even unto "snowball earth,"— have led to mass extinction events and the collapse of complex life into much simpler forms that can survive in extreme climate conditions. This explains why decline in solar activity leads to decline in biological growth processes as an adaptation to a lower energy environment. Our brains have evolved to synchronize our growth cycles to levels of solar activity.

The radical Eco-feminist movement that seems to dominate our culture today is hysterical about the human impact on nature. It is obsessed with energy conservation and a purported lack of environmental resources to support our growing populations and industrial-scale economies. This back-to-nature movement might actually be a biologically induced reaction rather than a rational, cognitive response to lower solar activity. It is likely a reaction that alerts our hormone regulators to halt our growth phase and go into hibernation mode. Diminished solar activity also leads to group conflict over the remaining available resources in a zero-sum game, in which no growth is possible.

In *The Testosterone Hypothesis* I analyzed how solar cycles drive our biological rhythm of life through hormonal cycles as the driving force behind the life cycles of complex organisms, whether as individuals or as groups of social animals that comprise entire civilizations. Here, I wish to further develop this thesis to include new information on how light–darkness cycles affect our activity and sleep cycles, sunspot cycles, and historical records of solar activity.

The cycles that affect life on earth begin with daily and seasonal hormonal cycles. The primary cycle in daily life is the circadian rhythm, in which the sun radiates light that regulates our daily hormonal patterns. In the morning, between 3:00 a.m. and 9:00 a.m., a gradual rise in solar radiation causes testosterone levels to rise, impelling us to wake up. During the daytime, testosterone levels reach a plateau, when we are the most active, and gradually decline into the night, causing us to become tired and fall asleep and bringing the daily cycle to its end. During the night the sleep hormone melatonin, "the hormone of darkness," made of serotonin and produced by the pineal gland in our brains, rises to regulate our sleep cycle. In many animals, including humans, a seasonal clock-effect is created by variations in the duration of melatonin secretion between summer and winter. The secretion duration at night affects the secretion profile of melatonin and signals the organization of seasonal functions that are day-length dependent, such as reproductive behavior.[2]

This hormonal mechanism, therefore, is responsible for translating variations in seasonal solar cycles into behavioral adaptations to the environment. As we shall further discuss, solar cycles vary not only on a seasonal basis, but also on much longer annual cycles, such as the 7-year and 70-year cycles. These are eras of mass social mood driven by recurring waves of testosterone rise and decline. The following brief section will aid in understanding the biological mechanism that acts as the coordinator to synchronize physical light-energy and the neuro-psychological motives that drive human activity.

The light–dark dichotomy is at the root of civilization ascendancy during periods such as the Enlightenment and periods of civilization collapse, like in the Dark Ages. Life on earth requires light energy to sustain and further itself, because the animal food chain is primarily dependent on plant consumption. Plant growth is in turn dependent on photosynthesis, which is the conversion of light into chemical energy stored in the form of carbohydrate molecules.

Periods of rising solar radiation levels signal our primitive mammalian brain that the environment accommodates a growth cycle, spurring testosterone, serotonin, and dopamine levels and propelling us to expand. In this cycle, we seek dominance in our own societies and over the rest of the natural world in order to acquire resources and reproduce. By contrast, periods of declining solar radiation impel us through decline in growth hormones to shrink into a mode similar to hibernation during winter in order to conserve resources and wait for a better opportunity to grow again. Adaptation to a changing environment is at the core of biological evolution. These hormonal cycles have been influencing human behavior for thousands of years, across the globe, by the great force that synchronizes our hormones: the sun. Light causes our master hormone regulator, the hypothalamus, to control the release of growth, sex, and stress hormones throughout our body.

## Our Brains and Light

In the opening lines of the Torah, Moses recorded the creation of the world, attributing to our benevolent Creator the first command of creation: "Let there be light." The scripture continues, "...and there was light. And God saw the light, that it was good: and God divided the light from the darkness." (Genesis 1:3, 4) Light is primal to life and to its very creation. It is no surprise, therefore, that light, and specifically bright light (more than 2,500 *lux*) has a direct effect on our ability to create life. Beginning in the 1980s, studies have confirmed that bright light exposure directly affects sex hormones. A 2003 study showed that when young men were exposed to one hour of bright light in the blue spectrum (BL), early in the morning for five days, the luteinizing hormone, a precursor to testosterone, increased by almost seventy percent.[3] As we know, improving testosterone production has many positive effects, including a decrease in depression and low sexual desire and function.[4] It is also possible that BL exposure in the early morning could stimulate ovulation in women, according to a 2007 study.[5] Notably, the color of light makes a significant difference in these and other studies. Light in the blue part of the spectrum is most effective when assessing testosterone production. Blue light has been shown to significantly increase the serum testosterone levels in rat babies when the mother was housed where predominantly blue rays affected her. The male rats were born with statistically significant higher levels of testosterone than those that had lived in white or black boxes. Green light also created higher testosterone levels in new-born rats, but not as high as the blue.[6] Likewise, blue and green lights positively affected testosterone production in young chickens, enhancing tissue fiber growth.[7]

Furthermore, ultraviolet-B sunlight rays trigger the production of vitamin D, which is a catalyst for testosterone and serotonin production. Vitamin D—a steroid hormone—is correlated with conception rates, both peaking in the same months in respective hemispheres. Peak sun exposure in the northern hemisphere in June and July leads to peak vitamin D levels in September and October, given the six to eight weeks it takes for the body to convert sunlight to hormone. Those are the months when testosterone levels are highest and when conception rates are highest.[v] An interesting exception to this pattern can be found in Scandinavia where sunlight is so pervasive in the early summer—twenty hours a day in June and July—that there is more of a bright-light effect, which causes the peak of conception rates to correlate with light exposure.[8]

Like testosterone, serotonin production is also light sensitive. It is considered the "oldest" neurotransmitter and is found in plants and animals in addition to humans. Sunlight's stimulation of plant photosynthesis also effects the serotonin levels in plants. Scientists have postulated that serotonin played an evolutionary role in bilateral animals. The "happy" effects of the hormone could have compelled a creature to continue seeking food in one direction or another with a positive feedback loop.[9] Serotonin is a known neurotransmitter in animal activities from swimming to brain development to socializing. In conjunction with antidepressants that affect serotonin production, light therapy has been used for decades to treat depression and anxiety, because of how light increases serotonin in humans. Whether it's Seasonal Affective Disorder (SAD) or non-seasonal depression, PMS or even suicidal depression, bright light therapy has been shown to aid people and decrease debilitating symptoms.[10]

In addition to specific light sensitivity, it seems that life on earth is sensitized to other facets of our sun's activities. The strength of earth's magnetic field, for example, or the actual position of the sun in the galaxy, or the occurrence of sunspots all play roles in creating and sustaining life. The earth's magnetic field, which is created by its core, protects the earth from solar particle storms by deflecting them into loops called the Van Allen belts. When solar radiation is particularly

---

[v] For a more comprehensive treatment of vitamin D and its effects, see my 2015 book, *The Testosterone Hypothesis*, Chapter 6, "The Solar Connection."

strong and the magnetic field is more highly activated, a section of the brain's hippocampus senses these changes and signals the pineal gland to produce melatonin and the hypothalamus to produce growth hormones. Scientist Maurice Cotterell argues for the *solar hormone theory* that asserts the sun controls fertility in women, coordinating its twenty-eight day rotation with women's twenty-eight day fertility cycle.[11] Could it be that the current obesity epidemic in the US is caused by low solar energy? Both testosterone and estrogen are involved in metabolic processes that affect, for example, glucose intolerance in aging women, leading to fat gain.[12][13]

Yet not only menopausal women are gaining weight in our culture, obesity is a problem across all age groups. At the same time, there are far too many cases of anorexism among women. It seems indicative of diminished sex hormone production that more and more Western women skew to extremes in body size, from anorexic to obese. This can been seen in the cultural icons of feminine beauty as well. When solar energy was high in the 1950s, the sex symbol of Marilyn Monroe had a voluptuous and curvy figure. Modern-day models beginning with Twiggy in the late 1960s, are extreme thin and even anorexic looking, which correlates with the decline in solar activity in the same years.

These extremes are definitive factors, among others, in the rise of infertility. We already know that men's lower testosterone levels have diminished sperm counts and lowered sex drive; and in conjunction with greater female infertility, it is no wonder there is a negative birth rate throughout Western societies. While other factors must be taken in to account, such as endocrine-disrupting chemicals in our water and food supply, our household products, and manufacturing materials, the historic trends and cycles of fertility and thus hormone synchronization present a picture much more comprehensive that just modern-day toxins could account for.

# 7 • SOLAR CYCLES

## Galactic Activities

Another dynamic of the sun's activity impacting life is its location in the galaxy. There are various theories about how this movement impacts earth's biodiversity, but the periodic alignment of galactic position and mass extinctions cannot be denied. Our solar system travels throughout the Milky Way in cycles of about 250 million years (Myr), and every 62 Myr, it completes a portion of that cycle. That is the time it takes for the sun, with all its planets in tow, to move and up and down in the galactic plane. This time frame is "suspiciously similar to earth's biodiversity cycle."[14] According to researchers at UCal Berkley and the University of Kansas, the fossil record indicates that approximately every 62 Myr, there has been a mass extinction of species groups (i.e., genera). Not all genera are apparently affected by the cycle, such as fish, squid, and snails; but other ancient creatures, like corals, sponges, and trilobites, do follow this pattern, and especially those species considered "short-lived," which survive for a maximum of 45 Myr.[15] A possible explanation for the galactic activity that could cause the extinction events is offered by Kansas researchers Adrian Melott and Mikhail Medvedev. Referring to the solar system's upward movement out of the crowded center of the galaxy, Melott proposes, "When we emerge out of the disk, we have less protection, so we become exposed to many more cosmic rays." The Milky Way's magnetic field is weaker when we are in that position, hence making the solar system more susceptible to destructive galactic wind.[16]

Mass extinctions are also evidenced in 30-Myr cycles and have been correlated with massive comet collisions with earth. A well-known incidence of this is the enormous crater found in Mexico in 1990 that is believed to have been the impact causing dinosaur extinction 66 Myr ago. Some 150 other craters have been found on the earth, and the geological record dates them at the same times of mass extinctions of species. The 30-Myr periodicity of such cosmic assaults are postulated to result from our location in the galaxy. One can readily see that this time frame is approximately half of the 62-Myr cycle, which puts our solar system in the midst of the galaxy at the 30-

Myr mark. What happens then is described well at the Oxford University Press "Daily Galaxy" blog:

> Normally, comets orbit the Sun at the edge of the Solar System, very far from the Earth. But when the Solar System passes through the crowded disc, the combined gravitational pull of visible stars, interstellar clouds, and invisible dark matter disturbs the comets and sends some of them on alternate paths, sometimes crossing the Earth's orbit where they can collide with the planet.[17]

When we look at perhaps the best-known mass extinction, the dinosaurs, scientists have found another tale related to solar activity. Dinosaurs evolved with *temperature-dependent sex determination* (TSD) as opposed to *genetic sex determination* (GSD). Like certain other reptiles—crocodiles, various turtles and tortoises, and some lizards—dinosaurs were potentially more susceptible to an offset ratio of males to females when the planet experienced significant temperature change.[18] As one *Smithsonian* headline queries, "Did Dinosaurs Die Out because Males Couldn't Find a Date?"[19] If the enormous pulse of a comet collision with earth caused a dramatic temperature change, and that climate shock caused many more males than females in creatures using TSD, the result might well have been "rapid reproductive failure."[20]

## Solar Cycles

Recognizing that solar radiation cycles of light and darkness affect life on earth raises the question of how these dynamics might influence or even govern the rise and fall of human civilizations over the past four millennia. To assess a pattern of the sun's activity with events on earth, we are indebted to many scientists who studied the sun's characteristics throughout the centuries. German astronomer Samuel Schwabe in 1843 looked at records detailing the numbers of sunspots and determined that there was an eleven-year period of variation. The number of sunspots cycled between a maximum and a minimum every eleven years.[21] In 1899, William Ellis ascertained that during every other minimum, there were "high counts of geomagnetically quiet days."[22] Eventually, these 22-year cycles were named Hale Cycles, and it was discovered that the odd and even 11-year cycles differed in the flow of cosmic rays toward earth. In reality the 22-year Hale Cycle is usually closer to 23 years, and close to half of that is the 11.5-year solar cycle. The approximately 22-year Hale Cycle of solar radiation explains the approximately 7-year cycle in social mood, as 3 times 7.5 is 22.5, close to 23. In addition, the 70-year cycle in social mood is a product of 3 times the 23 year cycle, which is 69, and almost 70 years.

These various cycles proved significant for those researchers who would consider specific qualities of the sun–earth relationship. The interdisciplinary study of the sun's effects on life and biological rhythms was first undertaken by Russian biophysicist Alexander Chizhevsky (1897 – 1964). He called this *heliobiology*, and its historical application is termed *historiometry*. Historiometry links the 11-year solar cycle with earth's climate and historical research of its effects on mass human behavior. Analyzing sunspot records, he compared them to the historical periodicity of riots, revolutions, and wars around the world for the period of over two millennia from 500 BCE to 1922 CE. Chizhevsky discovered that significant historical events correlated with recurring periods of sunspot maxima. He constructed an Index of Mass Human Excitability based on his studies. The Foundation for the Study of Cycles, established by Edward R. Dewey in the United States, published an analysis of Chizhevsky's data in 1951. In his 1971 book, *Cycles: The Mysterious Forces that Trigger Events*, Dewey further divided Chizhevsky's 11-year cycle into four parts:

1) Three-year period: has a minimum of excitability and is characterized by peace, tolerance, passiveness, lack of unity, and autocratic rule by minorities.

2) Two-year period: excitability grows. The masses begin to organize under new, revolutionary leaders. They challenge political and military decisions with new concepts, usually centering around one theme and encouraged by the press. Various leaders arise in local

situations with no apparent loyalty or alliance uniting them with other pockets of unrest.

3) Three-year period: is one of maximum excitability, but one which solves the most pressing problems of the era. Often these outstanding achievements are accompanied by the strangest insanities. Revolutions and wars abound, splinter groups congeal under one hypnotic leader, great military, political, and spiritual leaders emerge, and the people's voice is heard. The masses riot, bloody conflicts are commonplace, and the old establishment, in a most paralyzed state of inaction, offers only feeble resistance and is destroyed. Anarchy prevails but the fruits of its action are democratic and social reforms.

4) Three-year period: witnesses a gradual decrease in excitability until the masses are almost inert and apathetic. "Peace" is their cry, and their unity, so evident during the struggle, disappears. They slumber, like a bear in hibernation, coma like, awaiting a new season, a new cycle.[23]

Dewey's description of such a social cycle is reminiscent of the social cycles described in chapter 3, Neil Howe and William Strauss' generational turnings. Those authors argue for a generational cycle that spans more like twenty years, but many characteristics of their turnings can be seen in Dewey's cycles.

In the balance of this chapter, these patterns of social cycles will be considered using two different lenses. Relying on Maurice Cotterell's solar hormone theory, history can be reviewed through one lens. Conjointly, utilizing sunspot cycles, we can recognize similar and even more particular social phases.

In my last book, *The Testosterone Hypothesis*, I used global temperature charts as a proxy for solar radiation changes. Since then I discovered Maurice Cotterell's work using information on past solar activity to chart the rise and fall of civilizations based on *fertility cycles*. Cotterell utilizes the chart published by Professor Iain Nicolson in the book *The Sun* (1982), based on data collected by astronomer John Eddy from tree rings of Bristlecone pines that live up to four-thousand years. (Well preserved dead trees can even supply older data.) Nicolson was first to note that Eddy's graph "seems to follow the rise and fall of civilizations throughout history."[24] Cotterell's solar activity chart (Figure 7)[25] is quite similar to the global temperature chart of the last four millennia, which suggests a strong causal relationship between the level of solar radiation and earth's climate. However, because our master hormone regulator, the hypothalamus, releases hormones according to levels of light, it could be more accurate to use the solar activity chart to estimate past societal growth hormones and testosterone levels.

## Solar activity and the rise and fall of civilizations

Graphs showing that the rise and fall of civilizations corresponds with the rise and fall of radiation from the Sun; the graph shows a long-term envelope of solar magnetic activity (sunspots) derived from historical tree-ring data. Reduced solar radiation results in reduced fertility on Earth; the eighteenth dynasty of the sun-worshipping Tutankhamun collapsed during a massive minimum, as did the civilization of the sun-worshipping Maya 2,000 years later. The Inca city of Machu Picchu, occupied by the Virgins of the Sun from around A.D.1450, was abandoned when the Sun's radiation returned to normal levels in around A.D.1520.

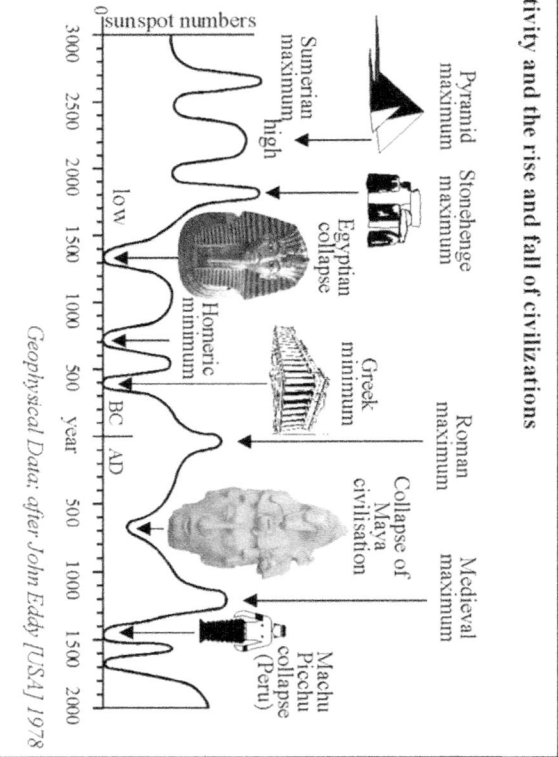

sunspot numbers

Pyramid maximum

Stonehenge maximum

Sumerian maximum high

Egyptian collapse

low

Greek minimum

Homeric minimum

Roman maximum

Medieval maximum

Collapse of Maya civilisation

Machu Picchu collapse (Peru)

3000  2500  2000  1500  1000  500  BC | AD  500  1000  1500  2000  year

*Geophysical Data: after John Eddy [USA] 1978*

FIGURE 7: Solar Activity and the Rise and Fall of Civilizations. (Courtesy of Maurice Cotterell, author of Future Science.)

It is apparent from Figure 7 that solar activity peaked during the Sumerian civilization in approximately 3300 BCE. As archaeologists have discovered, this was the most advanced culture of its time, inventing writing, the wheel, complex mathematics, an advanced legal system, and irrigation for agriculture. According to the Bible, Abraham, the father of monotheism, emerged from Ur, which was the most developed city-state of ancient Sumer. The rise of the peak-testosterone monotheistic worldview began with Abraham around the eighteenth century BCE and continued to evolve later in ancient Egypt and the land of Israel. As Figure 7 displays, the fall of solar activity over a period of four centuries, up to the fourteenth century BCE, led to falling testosterone levels and the culture of bondage, including the enslavement of the ancient Israelites by the Pharaoh in Egypt. This also correlates with the rise of transgenderism as discovered by archeologists. There is evidence that both Egyptian kings who reigned during the fourteenth century BCE, Akhenaten and his son Tutankhaten (i.e., King Tut), were androgynous.[26] Based on computer modeling of the body found in King Tut's tomb, it seems that this Pharaoh had an hourglass kind of figure and even small, female-like breasts. One account claims that Tut was "the epitome of androgynous royal glamour."[27]

The Jewish Exodus from Egypt is dated to approximately the thirteenth century BCE, a period in which solar radiation began to rise sharply again, leading to the return of the high-testosterone, masculine culture that empowers men to seek liberty and self-determination, in rebellion against tyrants. The kingdom of Israel peaked around 1000 BCE, aligning with peak solar activity, and then began to decline gradually, splitting in two in 930 BCE, with the northern kingdom falling to the Assyrians in 730 BCE and the Kingdom of Judea falling to the Babylonians in around 590 BCE.

This period of low testosterone, which is labeled on Cotterell's chart (Figure 7) as the Homeric and Greek minimum, led to the rise of homosexual Greek culture. As I discuss in *The Testosterone Hypothesis*, testosterone seems to have bottomed out and begun a rising trend, driven by increasing solar radiation, from 380 BCE. This change correlates with Plato's original praise of homosexual pederasty in the *Symposium* (385-370 BCE), to his condemning homosexuality in his final book, *The Laws*, in 340 BCE.

Solar activity and testosterone levels peaked around the birth of Jesus, and began a long period of decline that lasted until around 700 CE. This parallels the rise of Christianity under the Roman Empire. Christianity sought to unify humanity with a synthesis of its two parent cultures, ancient Israel and ancient Greece. The failed struggle for independence of the once great nation of Israel led to its destruction in 70 CE, with the fall of the Second Temple in Jerusalem. The ideal of Christianity gradually became altruism and sacrifice, rather than power and independence. Its pivotal events were the blood sacrifice of Jesus on the cross, a Jewish rabbi from Nazareth, and the belief that he was resurrected from death in order to achieve the communal goal of unifying the collective body of mankind. These events compelled his followers to promulgate their convictions throughout the Roman Empire, eventually developing the religion of Christianity that would later lead to the transmission of ancient Jewish wisdom and Greek philosophy into the nations of the world.

During the fourth and fifth centuries, the Christian theologian Saint Augustine formed the doctrine of Original Sin, equating sex and procreation with the sins of the flesh and the devil. In philosophy this period gave rise to extreme otherworldliness, known as Gnosticism. Gnostics sought to escape the dire and painful reality of this world to another dimension, separating the soul from the body, and idealizing the concept of an afterlife, where peace and quiet can be found, separate from the life of sin on earth. The concurrent decline of masculinity led to diminished family values and fertility rates and rising corruption by the centralized power in Rome, the capital of the Roman Empire. In addition, as the Roman legions weakened from lack of personnel, Rome relied more heavily on barbarian armies, and ultimately the empire collapsed into the barbarian chaos of the Dark Ages.

During the seventh century, the nadir of the Dark Ages, the Islamic cult of submission emerged from Arab desert nomads. It spread rapidly, conquering a huge empire across Asia and North Africa and reaching Europe through Spain. However, testosterone levels started a rising trend again in the eighth century, driven by rising solar radiation, and in 732 CE, King Charles Martel achieved victory in the Battle of Tours in France and stopped the advance of Islam into Europe. By 800, Charlemagne united much of Europe under the Holy Roman Empire and spread Christianity through the barbarian, pagan tribes. The rise in solar activity continued for five centuries until it peaked around 1200, leading to the Golden Age of Islam, a period of great scientific advance and rise of cultural learning centers in Bagdad, Alexandria, and Andalusia (Muslim Spain). Also, in Christian Europe, the population increased dramatically, with commerce and universities flourished during this period known as the Middle Ages.

However, solar activity began a declining trend between 1200 and 1450, again leading to the fall of great civilizations. The barbarian Mongol hoards conquered a huge land empire across Eurasia during the 1200s, decimating the great cultural centers of China in the East, Islam in the Middle East, and Russia in Eastern Europe. Gnosticism rose again in Europe, bringing the dualistic faith of Catharism, which was persecuted by the Catholic Church as heresy as part of the religious wars and the crusades in the period. In the 1300s a global cooling period began called the Little Ice Age. In this period, agricultural yields waned, and the Black Plague, which began in Asia, was spread by the Mongols' conquest. The Plague reached Europe in the mid-1300s, wiping out more than a third of the population in a few short years. The lowest level of solar activity, around 1452, caused the fall of the Byzantine Empire to Islam. However, a renewed rising trend in solar activity after 1453 promoted the information boom from the mass use of the Gutenberg printing press. This served to spread biblical literacy and ultimately spurred on the Protestant Reformation during the sixteenth century. Furthermore, in the late 1400s, the Age of Exploration began with the Portuguese quest for a route to the Far East; and within decades all the European sea-faring nations had staked a claim in the New World. However, another periodic collapse of solar radiation led to the Thirty Years' War during the period 1618-

1648. These years were followed by a long trend of rising solar activity that ushered in the Enlightenment and the scientific and industrial revolutions of the modern age.

One particular question of great interest is why the Judeo-Christian culture built Western Civilization, while Islam fell back to the Dark Ages in the twelfth century? The foundational difference between these two monotheistic religions, Islam and Christianity, is that Christianity is based on the peak-testosterone religion of the Hebrew Bible (Old Testament) that promotes reason and independence, while Islam is based on the Koran, a system of consummate submission to authority and group-think. The Koran was written during the trough of the Dark Ages by the prophet Muhammad (570 - 632 CE) and expanded its empire during the seventh century, reaching its peak in 750. As previously stated, when solar radiation increases global warming, testosterone and other happy chemicals in the brain increase. Mankind is then motivated to reason and develop the optimistic cultures, such as the Enlightenment philosophies—literally to be illuminated by light. As seen on the sunspot cycles chart below[28] (Figure 8), the culture of the Hebrew Bible was developed in a period of peak global warming, as was the modern West during the last four centuries of global warming.

During the crucial period of the 1680s, the last Islamic attempt to conquer the Christian West failed at the Siege of Vienna in 1683. This date was in the midst of what is termed the Maunder Minimum, identified in a 1976 paper by John Eddy in which he recognized the seventy-year period between 1645 and 1715 as a time when solar activity nearly came to a complete halt.[29] At this point, the West was emerging with a new vitality as solar radiation ended its down cycle, and began to rise. When global solar radiation rises, testosterone levels increase, at that time spurring the rise of the Protestant forces that returned to the high-testosterone philosophy of liberty inherited from the Old Testament. This process of cultural evolution is similar to epigenetic transformations in biological evolution that trigger new behavioral adaptations in the animal kingdom. The Glorious Revolution in England in 1688 and the Bill of Rights in 1689 occurred at the time the solar upswing was beginning to move the world out of the Maunder Minimum. These events guaranteed more political and religious liberty, particularly to Protestant sects, as the power of Catholicism continued to decline. Christianity thus rose up to build the West, while Islam never really recovered from its fall in the late middle ages. This might have been due to its Islam origins in the Dark Ages, which were dark because of a collapse in solar radiation that caused global cooling, and thereby a culture of depression in a perpetual state of holy war—or jihad—for global submission of humanity. Geography might have also have played a crucial role. Because of its distance from the East, the West was spared from the devastating wave of conquest by the Mongolian hordes that wreaked havoc on the Islamic caliphate. Similarly, the Mongols conquered much of Russia and subjugated it to a degree for the next three centuries. This could explain why the Russians never fully flourished in the capitalist system that the West did and why it fell into submission to communism in the twentieth century.

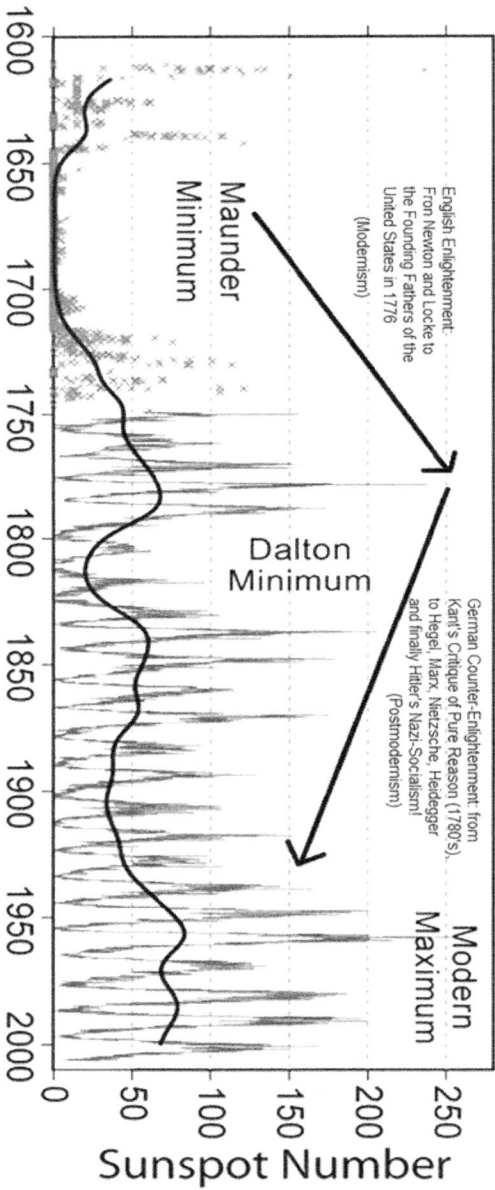

FIGURE 8: 400 Years of Sunspot Observations

# 7 • SOLAR CYCLES

## Sunspot Cycles

Having looked at Cotterell's schema of the solar-hormone connection, we turn to viewing human history in comparison with the cycles of sunspots. A direct measure of solar activity is indicated by trends in the quantity of sunspots, which have been measured for the past four hundred years. As can be seen in Figure 8, the episodes of low solar activity are called "grand minima" and have been given the names of prominent solar observers who spotted them. The coldest period of the Little Ice Age in Northern Europe in the seventeenth century is called the Maunder Minimum, and the less severe Dalton Minimum occurred between around 1790 and 1830. The overall picture of solar activity throughout the last four centuries, after a sharp decline into the 1650s, is a strongly rising trend, with great peaks in the 1770s. This was the period of the United States' Declaration of Independence. Another great peak, the Modern Maximum, is seen in ascendency from the 1940s into the 1950s, which led to the defeat of Nazism and fascism during World War II and a time of growing prosperity and resistance to Communism during the Cold War in the 1950s.

Marked with arrows on Figure 8, rising solar activity spurred the period of the Enlightenment, particularly by Puritans in England: from the natural philosophy of Newton and John Locke's ideal of liberty, to the Founding Fathers of the United States in 1776. But, after solar radiation peaked, it began a long period of decline from the 1780s to 1930s. This period of Modernism in philosophy eventually gave way to postmodern philosophy in the mid-twentieth century, the path to which included the German counter-Enlightenment, led by Kant's *Critique of Pure Reason* (1781), to Hegel, Marx, Nietzsche's existentialist philosophy, Heidegger, and finally culminating in Hitler's cult of Nazi-Socialism.[vi]

---

[vi] For a comprehensive discussion of these trends in philosophy, see my 2014 book, *The Objective Bible*, Chapter 3 "Pantheism."

# Sex Wars

## The 7-year and 70-year Cycles

The 7-year cycle of cultural highs and lows was observed in ancient times and is specifically described in the Bible. Joseph, son of Isaac, well known for being sold into slavery by his jealous brothers and rising to esteem under Pharaoh, interpreted a dream of Pharaoh's. In the dream was depicted a period of seven good and prosperous years followed by seven bad years of famine. Seven years is also given in the Torah, in the book of the law, Leviticus, as a year of rest. *Shemitah* years are 7-year time intervals that bring great historical upheaval to the financial, economic, social, and political spheres, The unfolding of events over the modern history of the recent century also validates this cyclical approach as the driving force in human civilization. In 2014, Messianic Rabbi Jonathan Cahn published *The Mystery of the Shemitah: The 3,000-Year-Old Mystery That Holds the Secret of America's Future, the World's Future, and Your Future!*. In the book, he lays out the five keys of Shemitah. Of particular interest here is the third key, which is the prophetic sign. In addition to the Sabbath year of rest and strict religious observance, Rabbi Cahn believes the seventh year offers portents of grave national issues. In the case of ancient Israel, the national ruin of 586 BCE that sent the people into the Babylonian exile was not only the destruction of the first Temple, it foreshadowed the destruction of the second Temple seventy years later in 516 BCE.[30]

The following list shows the Shemitah as demonstrated in the history of the past century:

1916-1917 – 40% US stock market value falls. The collapse of the German, Austro-Hungarian, Russian, and Ottoman Empires. The British Empire came close to insolvency during WWI.

1923-1924 – Weimer Germany's currency is wiped out in hyperinflation. In 1923 the Institute for Social Research, called the Frankfurt School, was established in Germany to spread cultural Marxism. The term *gay* was coined by lesbian Gertrude Stein in 1922.

# 7 • SOLAR CYCLES

1930-1931 – Global stock market collapse leads to the Great Depression. The Japanese invasion of Manchuria was in September 1931.

1937-1938 – 50% US stock market collapse caused another global Depression. *Kristallnacht*, a pogrom against Jews throughout Nazi Germany in November 1938, is carried out by SA paramilitary forces and German civilians.

1944-1945 – Defeat of Nazi Germany in August 1945 and the rise of the United States as the leader of the free world.

1951-1952 – Period of the Korean War, ending in 1953, led to the liberation of the southern nation from communism by the armed forces of the United States. US Senator McCarthy fights communism.

1958-1959 – Strongest-ever solar maximum recorded in February 1958. The Recession of 1958, also called the Eisenhower Recession, lasted for only eight months, followed by strong economic growth.

1965-1966 – 23% US stock market crash and recession, leading to the Six Days War in the Middle East in 1967.

1972-1973 – Almost half of US stock market value collapses, causing a global recession. The United States legalizes abortion and lost its first war, the Vietnam War. The 1972 Watergate scandal culminated in the resignation of US President Richard Nixon in 1974. In the Middle East the *Yom Kippur War* in October 1973.

1979-1980 – A global recession that led to the Islamic Revolution in Iran.

1986-1987 – 33% US stock market collapse, leading to recession.

1993-1994 – Global bond market crashes.

2000-2001 – 37% US stock market decline, the terror attack of 9/11, and global recession.

2007-2008 – 50% US stock market crash leads to global recession.
2014-2015 – Global financial collapse hits emerging markets and commodity markets. Russian military conflict in Ukraine, and the rise of ISIS in the midst of the devastating Syrian Civil War. Tensions between the United States and Russia escalate to almost Cold War levels.

## The 70-year cycle:

This great cycle has marked significant peaks and bottoms in history. In descending order, we can see:
1999 – Stock market top.
1929 – Stock market top.
1859 – Stock market bottom.
1789 – Recession leading to the French Revolution and United States Constitution and Bill of Rights.
1719 – Stock market Mania in France, the Mississippi Bubble, and the South Sea Bubble in England.
1649 – End of 30 Years' War in Europe. Treaty of Westphalia marks the rise of Protestant Christianity as the major force in Europe.
1579 – Protestant Dutch Revolt against Catholic Spain. The Northern provinces become independent, practically speaking, first in 1581, and legally in 1648.
1513 – *The Prince* written by Italian diplomat and political theorist Niccolò Machiavelli. This political philosophy written for the corrupt, tyrannical rule of the princes during the period of Renaissance Italy.
1453 – Invention of the printing press by Gutenberg in 1439 becomes commercial. End of the Hundred Years' War in Europe and the fall of the Byzantine Empire to Islam.

1383 – Wycliffe, heralded as the "Morning star of the
Reformation," was also an advocate for translation of
the Bible into the vernacular in the year 1382.

1313 – Dante Alighieri's *The Divine Comedy* is begun in
1308 and completed 1320.

*Prior 70-year cycles in history in ascending order*

1000 BCE – King David creates the great Kingdom of
Israel.

931 BCE – Split of Kingdom of Israel into two separate
kingdoms.

860 BCE – King Ahab follows Jezebel in the use of
witchcraft, idolatry, and child sacrifice.

790 BCE – Beginning of the period of Archaic Greece
with the rise of city states ruled by tyrants.

720 BCE – Fall of the Kingdom of Israel to the
Assyrians.

650 BCE – Sparta rises to become the dominant military
land-power in ancient Greece.

586 BCE – First Temple is destroyed by the
Babylonians.

516 BCE – Second Temple is completed under the
Persian King Cyrus the Great, 70 years after it was
destroyed. Start of the period of Classical Greece and
Roman Republic.

450 BCE – Temple and Statue of Zeus built at Olympia.

380 BCE – Plato writes *The Republic*

310 BCE – End of the period of Classical Greece and
beginning of Hellenism.

230 BCE – Archimedes produces great works in
mathematics such as his Treaties on Spirals.

167 BCE – Maccabean Revolt, a Jewish rebellion, lasting from 167 to 160 BC, led by the Maccabees against the Hellenistic influence on Jewish life.

110 BCE – The Hasmonean dynasty in Judea created the fully independent Kingdom of Israel.

63 BCE – Beginning of Roman Judea, Judea gradually losing its independence to the Roman Empire (emerged in 27 BCE).

0 – Jesus is born; Judea becomes part of the Roman Empire in the year 6 CE.

70 CE – Third Temple falls to the Roman Empire.

138 CE – The Bar Kokhba revolt fails, and the Jews are expelled from Israel, whose name was changed to Palestine by the Romans.

210 CE – End of Pax Romana, the period of Roman peace.

285 CE – Division of Roman Empire to western and eastern parts.

350 CE – Roman Civil War of 350–351 AD.

420 CE – Augustine writes *The City of God* in response to the decline of the Roman Empire with the sack of Rome by the Visigoths in 410.

476 CE – Fall of the Roman Empire.

542 CE – The Plague of Justinian afflicts the Eastern Roman Empire (Byzantine Empire). An estimated twenty-five million people die (more than a tenth of the world population at the time). Collapse in solar radiation causes extreme weather events during 535-536, an example of a volcanic winter, creating conditions for the plague.

610 CE – Muhammad starts preaching revelations publicly, proclaiming that "God is one," that complete surrender to him is the only way acceptable to God, and that he is a prophet and messenger of God.

680 CE – Battle of Karbala leads to split between Sunni and Shia Islam.

750 CE – Peak Islamic expansion during the Umayyad Caliphate (661–750).

820 CE – The Holy Roman Empire is established by Charlemagne.

890 CE – King Alfred the Great creates England.

960 CE – Otto I the Great crowned as Holy Roman Emperor.

1030 CE – The Golden age of Jewish culture in Spain ends with the fall of the Caliphate of Cordoba.

1099 CE – The First Crusade.

1170 CE – Return of Aristotelian philosophy in the works of Jewish Rabi Maimonides and Islamic Scholar Avicenna, which later influence the work of Catholic scholar Thomas Aquinas.

1240 CE – The barbarian Mongol Hoards conquer a huge empire, invading Kievan Rus, the old capital of Russia.

In addition to the 70-year cycles, it is instructive to see the events that transpire every 600 to 700 years, or in ten 70-year cycles. Table 2 shows events in descending time order from the twenty-first century CE to the twenty-third century BCE.

| Year | Climate Trend | Civilization Trend |
|---|---|---|
| 2000 CE | Peak modern warming period | Peak of the West |
| 1300 CE | Medieval Maximum | End of the Medieval Age, beginning of the Renaissance |
| 700 CE | Dark Age Global Cooling | The Trough of the Dark Ages. The religion of submission, Islam, conquers a world empire |
| 0 | Roman Maximum | Jesus is born and gives rise to Christianity |
| 586 BCE | Homeric Minimum leads to decreased solar activity | Second Temple is destroyed and fall of Judea |
| 1200 BCE | Rising solar activity | The rise of ancient Israel under the leadership Moses |
| 1800 BCE | Stonehenge Maximum | Abraham gives rise to monotheism |
| 2400 BCE | Pyramid Maximum | In the great Sumerian civilization, the "Epic of Gilgamesh" is written about the great king. This ancient poem is considered the first great work of literature |

Table 2: The 600-700 Year Cycle (Ten 70-year Cycles)

## How the 70-year and 7-year Cycles Drive Culture Wars

The period of 350 years, which amount to five 70-year periods, between 1649 and 1999, has been a fortunate one for the West. It has been a time of positive social mood that created the period of rationalism and advancement of the Enlightenment in which European society was transformed with the rise of reason, science, and liberty into the modern world. This was driven by high testosterone levels, caused by rising levels of solar radiation and global warming. The period just prior to the Enlightenment, the Renaissance, was another era of great flourishing; but, as presented in chapter 3, the prevailing energy of that time was more feminine.

1453-1525 was slated as the High Renaissance, in which the arts thrived in Italy. The period ended with the sacking of Rome in 1527. A remarkable figure of the period was Leonardo Da Vinci (1452-1519), who painted the famed androgynous figure of the Mona Lisa. At the period of the 1450s solar activity was very low, which probably caused many artists to be born homosexual due to low testosterone exposure, having a feminine brain type that is more artistic. However, solar radiation increased during this 70-year period, so that testosterone levels soared, leading to a burst of artistic creativity and the nullification of the Catholic celibacy vows by the German monk Martin Luther. In addition, testosterone mediates the rise in dopamine, which drives goal-oriented exploratory behavior, motivating Columbus in 1492 to sail west and discover the Americas, inaugurating a period of rapid European colonial expansion.

1525-1579 was a shorter period of rising solar radiation, causing religious wars, beginning with the Protestant Reformation started by Martin Luther in 1525, leading to the rise of Calvinism and separation from the centralized authoritarian model of the Catholic Church.

1579-1649 was a period ending in declining solar radiation, leading to the Thirty Years' War and upheaval in the European Reformation between the Protestants and Catholics, ending with the Protestant North gaining dominance.

1649-1719 occurred during the Maunder Minimum period of global cooling (1645-1715). The cold weather caused crop failures, and one theory contends that this could be what drove the hysteria of the 1692 witch trials in Salem.[31] Solar radiation bottomed out in the middle of this period, the 1680s, and started in a rising trend that would last for the coming centuries. In the 1680s both science and philosophy experienced great advances with the publication of

Newton's classical mechanics and John Locke's work on liberty and individual rights. [vii]

The remainder of this list is enhanced by the chart (Figure 9)[32] that depicts the annual number of sunspots observed and reported in the years following 1700.

---

[vii] A larger analysis of the correlation between European witch trials, weather trends and agricultural economy can be found in an intriguing article, "Witchcraft, Weather and Economic Growth in Renaissance Europe," by Emily Oster, published in the *Journal of Economic Perspectives* 18, no.1, (2004): 215–228, http://ergotism.info/en/oster_weater_witches.pdf.

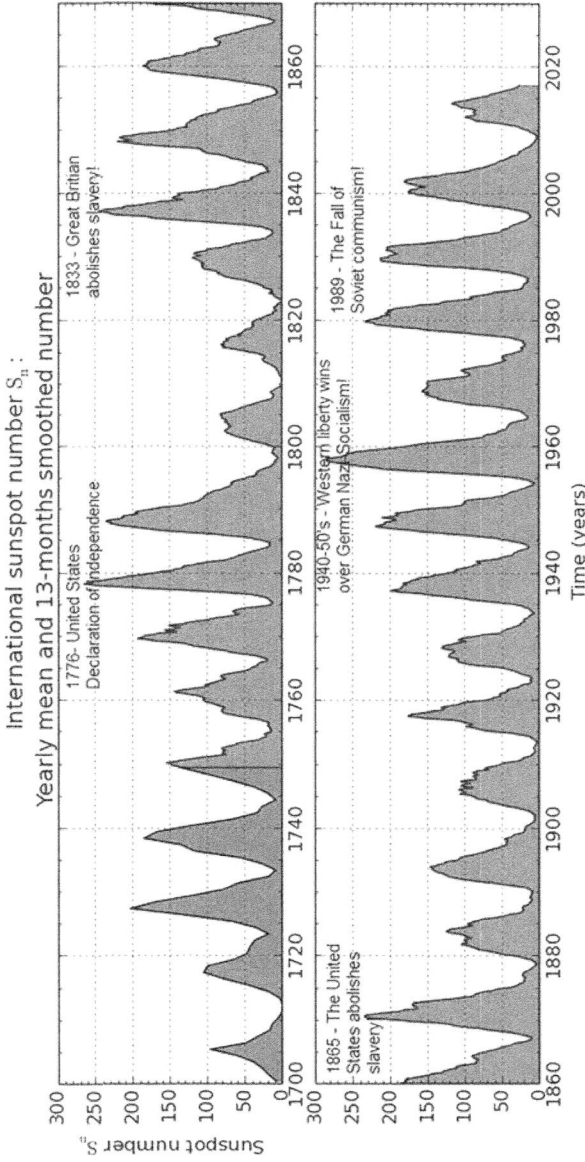

International sunspot number $S_n$ :
Yearly mean and 13-months smoothed number

1776- United States Declaration of Independence

1833 - Great Britian abolishes slavery!

1865 - The United States abolishes slavery

1940-50's - Western liberty wins over German Nazi Socialism!

1989 - The Fall of Soviet communism!

Time (years)

Sunspot number $S_n$

SILSO graphics (http://sidc.be/silso) Royal Observatory of Belgium 2017 October 2

FIGURE 9: Yearly Mean and 13-Month Smoothed Sunspot Number

1719-1789 – a period of rising solar radiation until about 1776, with an intermediate decline in the period 1740-1770, was ending in a large decline into the Dalton Minimum in the years 1780-1820.

This period began with huge financial bubbles and crashes during the 1720s: the South Sea Bubble in Britain and the Mississippi Bubble in France. Decline in social mood during the 1740s to 1770 led to David Hume's attack on reason as a slave to the passions and Rousseau's radical feminist-socialist egalitarian philosophy of back to nature, associated with his enchantment with the "noble savage." Rise in solar radiation during the 1770s led to the American Revolution by the Founding Fathers, with the masculine ideas of struggle for independence and liberty from tyrants.

*The modern world*

Between 1789 and 1999, 210 years, or three periods of 70 years, we were in a rising wave of social mood, driven by the rise in testosterone, primarily until the 1950s and gradually declining since then:

1789-1859 – This period began with global cooling during Dalton Minimum and falling solar activity until 1810, which resulted in French Revolution and the Napoleonic Wars ending in 1814. The significant publication of Kant's work attacking reason and reality in philosophy and the Marquis De Sade's promotion of Sadomasochism and malevolent sexual philosophy all occurred in this time frame.

As seen in Figure 9, there is a rise in solar radiation up to the late 1830s that correlates with a period of rising social mood and ends in a stock market bubble in 1837. In 1833 a peak positive-mood trend led to the British Empire's abolition of slavery. In the two decades of 1839-1859, fall in solar radiation led to a period of negative social mood, beginning in 1839 with a 4-year period of deflation and economic depression, liquidating the financial bubble caused by the previous euphoric period of economic speculation. In this period were the Opium Wars between the British and Chinese, development of Marxist and Darwinian theories, and a growing trend toward socialism.

# 7 • SOLAR CYCLES

1859-1929 – The previous deteriorating trend in social mood culminated in the American Civil War from 1861 to 1865, and the victory of the Republicans in the North with the final abolition of slavery in the United States. Sunspots peaked in the early 1870s, followed by a downtrend that lasted into the 1930s. This era instigated a gradual rise of feminism and socialism, leading to the Bolshevik revolution in Russia in 1917 and two world wars.

1929-1999 – Severe testosterone collapse caused the Great Depression in 1929-1933 and spurred the rise of communism, fascism, and National Socialism in Hitler's Germany, bringing WWII between 1939 and 1945. However, solar radiation cycles began to rise remarkably in 1942, and during the 1940s and 1950s, the defeat of fascism and the triumph of Western liberalism came about with the return of a high-testosterone culture. Again, sunspot cycles peaked in the late 1950s and fell into 1978. During this time the US fought and failed in Viet Nam, and the feminist sexual revolution prevailed. In China under Mao, the Cultural Revolution swept the country, and by the late 70s, the Islamic revolution ousted the Shaw in Iran, and the Soviets marched into Afghanistan.

In the 1980s sunspots rose again, leading to a positive period in social mood but to a lesser degree than in the 1950s. Nevertheless, testosterone levels were elevated sufficiently to bring the fall of the Soviet Empire that had lasted for seventy years, 1919 to 1989, from the time of last great decline in social mood. In the following chapter, the period following the 1980s will be reviewed, along with some more particular vantage points of viewing historical trends that are linked to solar activity, from volcanic eruptions to the stock market.

# 8

---◦◦---

# THE SUN, CLIMATE CHANGE, AND HUMAN CULTURE

One might never imagine that the tale of Frankenstein would have any connection to stock prices or, for that matter, a plight of locusts overcoming crops in Iowa. Yet, what these and other phenomena do have in common is the sun's activity. In ways more and less obvious, as illuminated in the previous chapter, solar radiation levels exhibit their effects on our earth.

In the summer of 1816, famed English poet Lord Byron hosted a group of artists and writers at his home in Switzerland. Their intent to spend days hiking the mountains and boating the lakes was thwarted by the freakish weather. It stormed and even snowed throughout the summer months; it was dark for days on end—even weeks. And so the guests were confined to the house. Finding themselves in this unexpected circumstance, they decided to entertain themselves by writing ghost stories. It was from this challenge that writer Mary Shelley composed the foundational story of the monster-man that would become one of the most popular and well-known sagas of the modern world: *Frankenstein*. If one is familiar with the novel, or even one of the films it inspired, it might be recalled that the weather garners great attention in the story. The storms, the dark cold, the ice and snow all remind a reader of the frigid years Shelley wrote this story.[1] Her host, Lord Byron, captured the ghastly weather of that summer with these words:

I had a dream, which was not all a dream.
The bright sun was extinguished, and the stars
Did wander darkling in the eternal space,
Rayless, and pathless, and the icy earth
Swung blind and blackening in the moonless air;
Morn came and went—and came, and brought no day,
And men forgot their passions in the dread
Of this their desolation; and all hearts
Were chilled into a selfish prayer for light.[2]

Why was the "bright sun extinguished?"

What had caused a weather event that cut the New England growing season in half and scourged European wheat yields by seventy-five percent? The catastrophe that was world-wide began with an incident in Indonesia in April 1815: the eruption of the volcano Mount Tambora, the largest volcanic eruption ever recorded in human history. By the summer of 1817, crop prices in the Alpen territories of Germany and its neighboring Switzerland were three-hundred percent higher than the year prior. The population "resorted to eating rats, cats, grass and straw as well as their own horses and watchdogs."[3] The eruption of Mount Tambora, experts report, occurred in a year of a solar minimum following a pattern of solar minima and increased volcanic activity. Periods of declining solar activity cause greater flux in cosmic radiation, which promotes more cloud cover, as well as greater volcanic activity that results in dust covering the atmosphere and subsequent global cooling.[4]

Many significant human events happen during the years surrounding large volcanic blows and the colder temperatures from sharp declines in solar radiation. Tree-rings show "a distinct temperature decrease during solar key years due to a coincidence between solar and volcanic forcing."[5] Grand solar minima were seen in 1282, 1458, 1698, and 1817. Within twenty-five years of these minima, there was a global increase in medium to large volcanic activity.[6] In some of those years, the following events took place: 1258, Islam fell to the Mongols; 1453, Byzantine fell to Islam; and 1815, the Battle of Waterloo ended the Napoleonic Wars.

The desperation from agricultural failure is one horror, but falling solar radiation sadly causes others. The decline in growth hormones and global temperatures and the drop in agricultural productivity as photosynthesis wanes all result in social conflict over available resources. This leads to further deterioration in living conditions and declines in general public health. Major epidemics subsequently break out, such as the Black Plague in the fourteenth century, the Spanish Flu epidemic following WWI in 1918, and the AIDs epidemic in the 1980s, following the rise in homosexuality during the 1970s.

The book *Black Death and Abrupt Earth Changes in the 14th Century* by Sacha Dobler depicts abrupt astrological, tectonic, and climate events in the period 1290-1350, leading to the epidemic that is estimated to have killed almost half the European population in a period of a few years. Earthquakes caused dust to cover the sky and exacerbated the global cooling. This created agricultural shortages, starvation, and wars, such as the Hundred Years' War between England and France (1337-1453). Social strife and conflict led the Mongols to attack the Italian trading city of Caffa in the Crimea in 1347, spreading the plague from Asia into Europe by fleeing Genoese traders.[7]

Yet another plight of, literally, biblical proportion is locust invasion. Before the Black Plague, there were years of locust invasions into Western Europe, particularly in 1336-1338, along with earthquakes, global cooling, lack of rainfall, and years without summer in the 1340s. People could not help but be reminded of the plagues in Egypt, as part of the ten calamities that God inflicted on Pharaoh around 1300 BC, triggering the Exodus of the ancient Israelites. Also during the Black Plague, Jews were blamed and persecuted in Europe as responsible for spreading the disease.[8]

As described in chapter 4, a locust swarm is a collective of grasshoppers, which, due to a lack of food from deteriorating environmental conditions, coalesce in swarms of millions in order to fly together into new territories to acquire food. This action is mediated by changes in serotonin levels, transforming the anti-social insects into a cohesive social structure. It is very probable that the same changes in environmental conditions, that is, falling levels of solar radiation, cause phase transitions in human social dynamics as social groups adhere together to fight for available resources under extreme stress. During periods of declining solar radiation and recession in 1915, 2004, and 2013, outbursts of locusts could be seen swarming over the desert in Israel, for example.

## The Sun and Stock Markets

Volcanoes, locusts, disease—all potentially calamitous happenings for life on earth. The consequences of regional or global food shortages and widespread public health crises are bound together in a complex web that includes the financial state of the globe. Long before fiber-optic cables beneath the oceans or satellites hovering in the heavens connected global communities, the world was linked through trade and immigration. And like the natural disasters just discussed, global financial markets show distinct patterns that correlate with solar activity.

Assessing solar minima in an April 2009 article, *Deep Solar Minimum*, NASA reported data that corroborates this correlation. Following the catastrophic global financial collapse that bottomed only a few months earlier, these significant observations on solar activity strongly suggest that the negative trend in social mood, caused by a sharp drop in solar radiation, led to the catastrophes of both WWI in the early twentieth century and the great depression of 2008 a century later. In 2008 no sunspots were seen for seventy-three percent of the year. Only in 1913 were so few sunspots reported. These lows in solar brightness could be measured in 2008 by NASA spacecraft that showed a 0.02% decrease since 1996 of visible wavelengths and 6% decrease in UV wavelengths. The article draws a parallel to the 1901-1913 "deep calm" solar minima.[9] Another indication of low solar activity leading to the 2008 financial crash was the fifty-year low in solar wind, also reported by NASA.[10] They measured a twenty percent decline in average pressure of solar wind in the same twelve-year period noted above, 1996-2008; and this power was at its weakest point since solar wind had first been recorded in 1959.[11] During this same time period, decline in solar radiation, causing falling serotonin levels, led to the rise of mass depression and a fourfold increase in the use of anti-depressants.[12]

*Hemlines and Stocks*

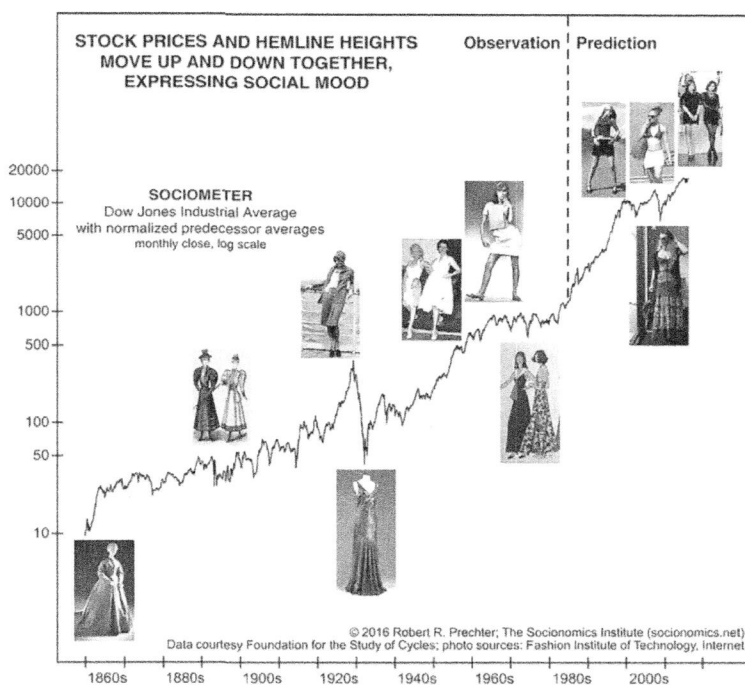

Figure 10: Hemlines and Stocks [13]

A somewhat amusing but no less true correlation with stock prices is fashion trends. Considered by some to be an urban legend, it turns out that hemlines really do rise and lower following the trends in prosperity levels. Figure 10 depicts the Dow Jones Industrial

Averages since 1860 and shows the relative hem lengths associated with market highs and lows. This is indicative of how sex hormones drive social mood as reflected by cultural and sexual norms in fashion trends. When solar radiation rises, such as in the early 1870s, and 1950s, social mood and testosterone levels improve, which triggers a more joyful and happy atmosphere, more open sexual attitudes and shorter hemlines. However, as solar radiation levels decline, leading to testosterone collapse, as in the 1850s, 1930s, 1970s, and the 2008 recession, periods of depressive sexual behavior lead to longer hemlines. In 2010, researchers at Erasmus University in the Netherlands made a prediction in this realm. Based on a three-year time delay from market peaks and troughs, they expected that ankle-length skirts would come into vogue in 2011 or 2012. And indeed, the "maxi-skirt" made a roaring comeback in those years, just as predicted.[14]

In chapter 7 the case for cultural fluctuations with solar cycles was presented. The 70-year cycles discussed can be seen in the stock market trends, as well. Ralph Nelson Elliott developed the Elliott Wave Theory in the 1930s, suggesting that stock markets trade in repeating cycles that are governed by recurring fractal patterns of social mood and behavior of mass psychology. According to this principle, which I explain in part three of *The Objective Bible*, we can study stock market, financial, and economic cycles as an indicator of social mood. The 70-year cycle can be seen in Figure 11, a socionomic chart of social mood trends (stock market trading data is only available from 1700).

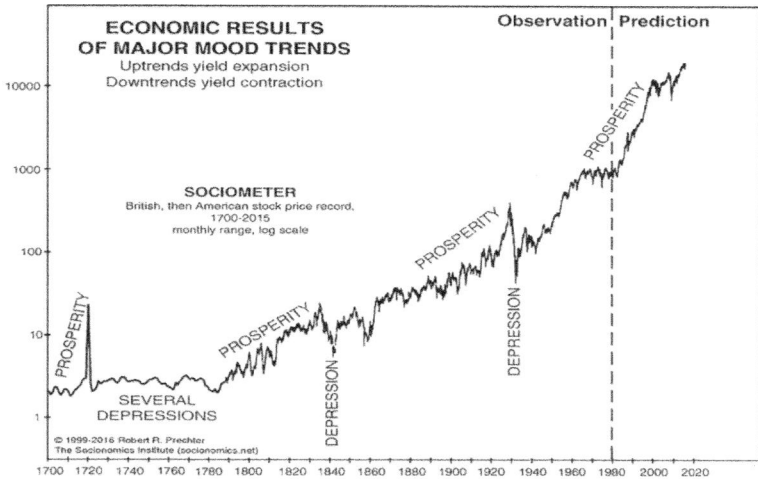

Figure 11: Stocks and Social Mood Trends, 1700-2017 [15]

1650-1720 is a period of an up-trend in social mood and prosperity ending in a bubble.

1720-1790 is a period of several depressions, ending in the American and French Revolutions.

The 210-year up-trend period can be divided into three 70-year phases:

1790-1860 is the first up-trend period until the late 1830s, ending in a prolong recession.

1860-1930 is a period of great economic growth, ending in the financial bubble of 1929.

1930-2000 is a period beginning in depression and WWII until 1945, but ending in great prosperity.

*The Sunspot Cycle and Stocks*

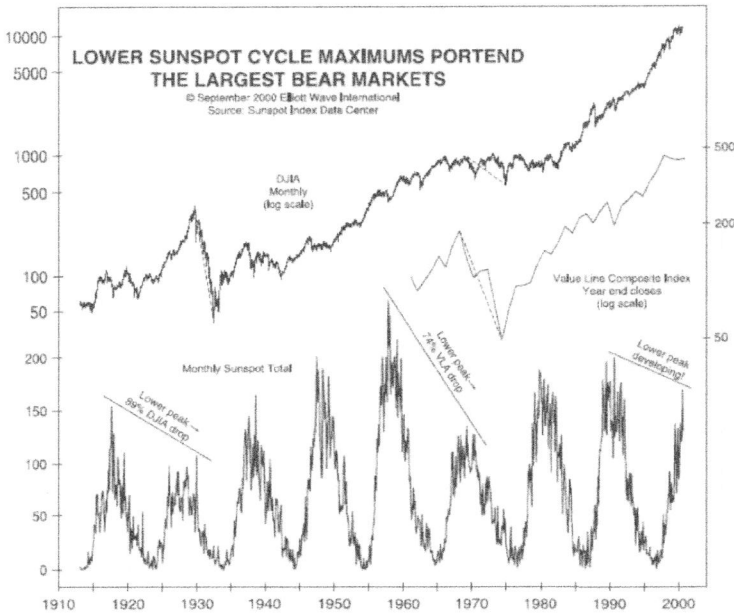

Figure 12: Sunspot Cycle Maximums and Dow Jones Industrial Average, 1912-2000 (Courtesy of Elliott Wave International, *The Sunspot Cycle and Stocks* (2000), Elliott Wave Theorist) [16]

Diverging behavior between solar activity and financial market trends is a warning signal of imminent trend change, as subtle transitions in hormone levels effect social mass behavior. This was used by Elliott Wave International to predict the coming stock market debacle of 2000, as the top of the stock market bubble was forming earlier in the year. Figure 12 shows the slope lines on the sunspot graphing that depicts the declining solar energy in the years prior to the 1929 and 1968 stock market highs. The correlative stock graphing above shows those highs and the ensuing bear markets, which were the worst of the twentieth century. Based on the activity of sunspot "bunching" in 2011 and the fact that the cycle was "topping out at a lower level than that of 1990," the Socionomics Institute considers that this might indicate a lower maximum of sunspots in the cycle, which is what can be seen in Figure 13.[17]

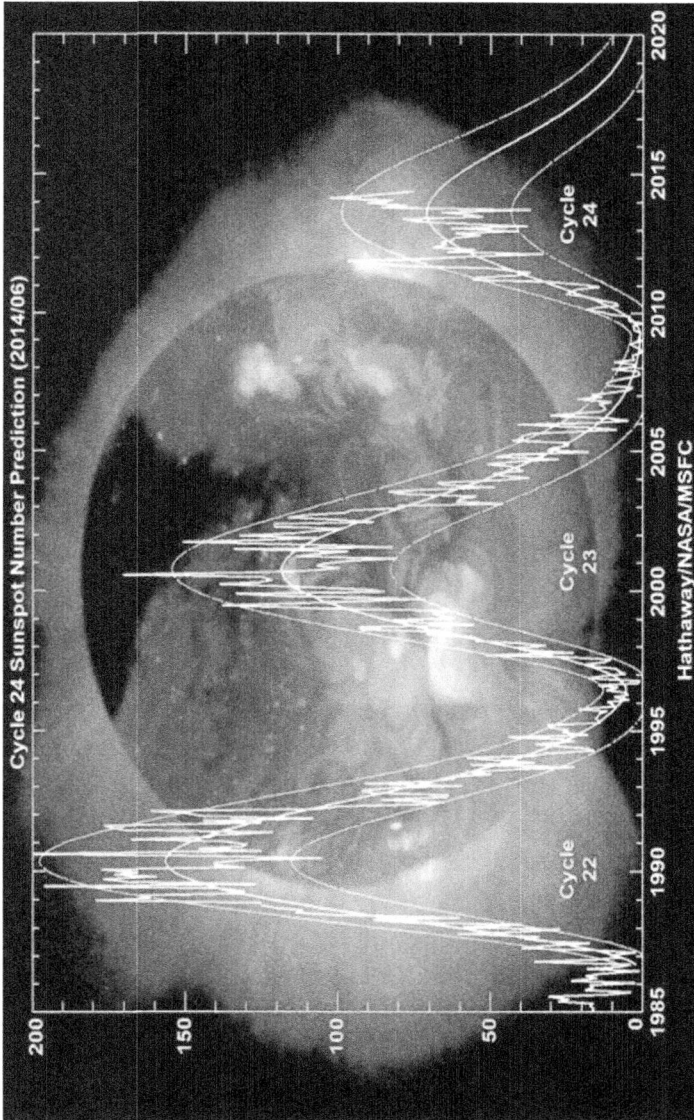

FIGURE 13: Sunspot Number Predictions

According to the Elliott Wave Theory, the next 70-year period we should expect will be one of negative social mood, similar to the period of 1720-1790. This negative mood trend should be driven by declining solar activity, predicted by scientists in *The Astrophysics Journal* of November 2014. They expect this decline based on the magnetic phase shifts occurring in the sun's magnetic poles that will subsequently further decreased sunspot activity. Their model indicates the next ten to twenty years of solar radiation decline, up to eighty percent in the current cycle and another forty percent by 2040.[18]

In financial markets, bull markets are the rising waves, while bear markets are the declining waves. The raging bull is also the symbol of the high-testosterone independent male who cannot be controlled and is often castrated in order to be domesticated and controlled by humans. The bear represents the winter hibernation period of severe decline in activity. Author John Coats has written *The Hour between Dog and Wolf: Risk Taking, Gut Feelings and the Biology of Boom and Bust* (2012) in which he looked at the behavior of stock traders during growth markets, especially bubbles. He found that they exhibited modes of testosterone level change similar to male animals that have been studied when they are competing for mates or territory. The winner of such fights not only benefits from the testosterone that "increases their blood's capacity to carry oxygen and, in time, their lean-muscle mass... [and] increases the animal's confidence and appetite for risk,"[19] he benefits from the elevated testosterone in ensuing battles. This is called the "winner effect," and could be the reason that athletes and gamblers and business people are capable of "winning streaks." There is a tipping point, however, when the push to risk goes into overdrive and "confidence and risk-taking segue into overconfidence and reckless behaviour."[20] Coates suggests that the winner effect in overdrive propelled stock traders in the 1990s as the technology bull market turned into a bubble, making traders "every bit as delusional, overconfident and risk-seeking as those animals venturing into the open, oblivious to all danger."[21] Arguably, the same behavior factored into the housing bubble that burst spectacularly in

October 2008, plunging the US and the globe into a deep recession for almost a decade.

The dynamics of markets and economies and social mood play out at various levels, micro and macro, as evidenced by the winner effect at stock exchanges in New York, London, Hong Kong, and elsewhere. These traders, most of whom are men, exist in a kind of micro-organism within the human superorganism, yet their propensities at times reflect and at times shape social mood that is enormously dependent on the hormonal activity modulated by the sun's energy levels. In another realm of the finance industry, technical analysts of the market essentially analyze various quantitative parameters to assess how social mood drives market prices. The market's directional movement is a measure of our society's collective optimism or pessimism about our economic growth and productivity.

There is a famous saying among analysts that a bull market climbs a wall of worry, while a bear market descends a wall of hope. The global financial collapse of 2008 was caused by incredible stress levels and a corresponding rise in cortisol. Then, as stress levels reached their peak in March 2009, the financial markets bottomed out, and testosterone began gradually to rise again. The markets began to climb higher, piercing through the wall of worry as confidence progressively returned to investors. However, testosterone levels never reached the levels considered normal in the period prior to their catastrophic plunge of 2008, and fear and anxiety remained remarkably high, as evidenced by the rise of social polarization, homosexuality, transgenderism, sadomasochism, feminism and Islamic jihad.

Because of a sharp downtrend in social mood and the culture of fear, the stock markets, particularly in the United States, have actually rebounded sharply, even reaching all-time highs, at least in nominal terms. This activity represents a "manic" market characterized by all-time low global interest rates—even lower than during the Great Depression and WWII in the 1940s cycle—leading investors to seek financial speculation in a scramble for yield. In 2017, although the financial markets have been rising since the bottom in 2009, testosterone levels seem to have dropped in the later part of the up-cycle, as the raging bull of Wall Street is now in a crude face-off with the feminist opposition. There is no better image of the current conflict in our culture that the "Fearless Girl" statue fiercely "staring down" the iconic bronze bull of Wall Street. The girl was put up in March 2017 as a protest against purported pay inequity and the male–female power gap in Wall Street firms.[22] Yet, the message conveyed does not exemplify women professionally joining the exuberance of bull-market prosperity; rather, it's an image of staunch opposition to what the bull signifies. The message, it seems, is confused. The four-foot tall girl communicates defiance that contradicts the call for increased engagement and opportunity for women on Wall Street. This confusion is part and parcel of the present state of psychosis caused by decreasing solar activity and tumbling testosterone levels in the culture. The market is now in a similar condition to a manic-depressive disorder, rising in a manic phase that will end in a bitter collapse and another great depression.

This kind of cultural bipolar condition indicates a rise of oxytocin and adrenaline, hormones that also elevate dopamine, which supplies the brain with intoxicating feelings of what former Federal Reserve Chairman Alan Greenspan in 1996 famously called "irrational exuberance." We can see in Figure 13 above that solar activity has been falling since the 1980s; and clearly symptomatic of this is the addictive profile of financial speculation as well as addiction to drugs, porn, sadomasochism, gambling, video games, sugar, and seeking instant gratification through casual sex rather than long term relationships—not to mention pervasive cultural anxiety and depression. By the year 2015 the state of affairs in the US was such that author David Kupelian wrote in *The Snapping of the American Mind*:

> According to the Centers for Disease Control and Prevention, 130 million Americans depend on some sort of mind-altering substance to get through life. Some 110 million Americans have sexually transmitted diseases. Families are disintegrating. One in four middle-aged women is taking antidepressants. We've become a nation of addicts.[23]

Because people are suffering from a negative social mood but do not understand the reason for their rising stress levels, they seek to alleviate their condition by artificially stimulating their happy chemicals to rise and produce short-term feelings of euphoria. Thus, the impetus to become addicted to various stimuli. However, as studies of addiction demonstrate, our brain has inherent mechanisms to increasingly adapt to new stimuli, and hence, the addicted brain requires ever-increasing amounts of stimuli to supply the same sensations of relief. The concentrated production of extracellular dopamine that addictive substances and behaviors ignite "may raise the thresholds required for dopamine cell activation and signaling."[24] Brain imaging has proven that drug addicts have fewer dopamine receptors and minimized dopamine activation:

> These findings implicate deficits in dopamine activity—
> inked with prefrontal and striatal deregulation—in the
> loss of control and compulsive drug intake that results
> when the addicted person takes the drugs or is exposed to
> conditioned cues. The decreased dopamine function in
> addicted individuals also reduces their sensitivity to
> natural reinforces.[25]

As the brain's dopamine receptors become less responsive and solar radiation continues to decline, the manic phase of euphoria begins to change into deep depressive mood. We have seen this manic–depressive behavior of social mood manifested in financial mania in periods such as the great stock market bubbles of 1929, 2007, or 2016. Falling testosterone levels are evident in the rise of transgenderism and decline in gender differentiation, but the population is still able to grasp onto some false hope by clinging to fantasies of prosperity driven by a debt-fueled consumption spree that will all but vaporize as the air blows out of the bubble.

An increase in altruism attests to the rise in the female bonding hormone oxytocin that leads to a greater trend toward collectivism; and high stress levels lead to escalating cortisol, the stress hormone, which elevates social animosity toward "others" and incites conflict. When the manic phase of the bubble reaches its spectacular end, it will inevitably bring depression, with financial and economic implosion and another Great Depression, leading to global war.

*The 7-year cycle in negative social mood*

Even with a period of overall rising trend in social mood, as in the 21-year period of the 1980s and 1990s, there was an intervening 7-year period of negative social mood, leading to an economic recession and gender and race wars. During the 1987-1994 period of negative social mood, we witnessed various manifestations of the culture wars:

*Sexual Personae: Art and Decadence from Nefertiti to Emily Dickinson* (1990): Camille Paglia's book describing the androgynous personality and sex wars as the primary driver of Western culture.

*The Adapted Mind: Evolutionary Psychology and the Generation of Culture* (1992) is considered to have given birth to modern evolutionary psychology, integrating evolutionary biology and cognitive psychology.

*Men Are from Mars, Women Are from Venus* (1992): John Gray's hugely popular book depicting the psychological differences between the genders that used the metaphor of different planets, each acclimated to particular social customs.

*The 1992 Los Angeles riots*: civil unrest and uprising amid racial violence following the Rodney King trial, in which white police were acquitted for the assault of Rodney King.

*Falling Down* (1993): a film depicting an unemployed divorced man whose life deteriorates into a cycle of violence, amid growing social strife.

*O. J. Simpson murder case*: in June 1994 a trial over the murder by Simpson of his ex-wife and her friend. The most publicized criminal trial, involving racial elements as Simpson was black and the victims were white.

*The Bell Curve: Intelligence and Class Structure in American Life* (1994): an academic book by political scientist Charles Murray and psychologist Richard J. Herrnstein, They argue that human intelligence is primarily determined by genetic factors, stating that class and racial differences are influenced by genetic inheritance. It met with vociferous resistance from the academic community.

*Seinfeld:* an American sitcom on NBC, which was featured from 1989 to 1998, and depicted the trend toward a nihilistic culture. The main characters have no families, children, or any particular interests or pursuits in life. This was followed by *Sex and the City:* a television series broadcast from 1998 until 2004, depicting unmarried women in New York City pursuing short-term sexual adventures, with no long-term relationships.

Another, more current example of sex wars that is symptomatic of the declining solar radiation and testosterone levels is the call for gender-neutral pronouns. This demand differs from earlier attempts to introduce a gender-neutral single pronoun into the English language. In the 1850s the terms *nis, nir, hiser,* and *thon* were suggested, as reported in an 1884 article in the *New-York Commercial Advertiser.*[26] Likewise in the 1930s, thon was circulated in another attempt to offer a singular, neutral pronoun in English to speak in generalities. Thon actually remained in *Merriam-Webster's Unabridged Dictionary* until 1961.[27] As a grammatical tool, it's logical enough to want a term for statements like "The students each took his or her books home," where "his or her" is clunky phrasing. Such a term exists for the plural in English, they or their. However, in the twenty-first century cultural climate, colleges like Harvard propose multiple options for personalized pronouns such as *se, ze, hir, zir,* and so on[28] and the city of New York has passed a law making it an illegal offense—with fines reaching a quarter-of-a-million dollars—to call people by pronouns other than their stated preference.[29] Actions like these are meant to untether a person from the "binary" male/female identity. They are indicative of the revolt against biological categories for gender and have been vigorously promoted by LGBTQ activists.

### Conclusion

Our culture is a product of psychological factors that originate in our biological evolution, experimenting with various forms of sexual behavior that determine our social organization in order to adapt to changing environmental conditions. Solar cycles, determined by the location of our solar system in the Milky Way galaxy and even the cosmos, is the primary driver of biological growth patterns on earth. Evolutionary psychology determines that natural selection has shaped our mind in order to adapt these energy flows that are required to sustain our life processes.

Our academic establishment, based on a reductionist method of science, has yet to comprehend this big picture view of our cultural evolution, failing to integrate the physical, biological and social sciences into a holistic view of our place in the universe. However, with the rise in recent decades of the new scientific view of nature as a complex adaptive system, we now stand at the cusp of a paradigm shift that will propel us forward in a scientific revolution. We are poised to relate with more comprehension and accuracy to the interconnected factors that shape our physical, biological, and social existence. Hence, in the next chapter we shall discuss progress in the science of physics since the Enlightenment and shed light on how our view of the physical world is shaped by sex hormones, which in turn, are regulated according to levels of solar activity.

# 9

# HOW SEX HORMONES SHAPE OUR VIEW OF THE PHYSICAL UNIVERSE

*We can speak and think only of what exists. And what*
*exists is uncreated and imperishable for it is whole and*
*unchanging and complete. It was not or nor shall be*
*different since it is now, all at once, one and continuous.*
— Parmenides
*No man ever steps in the same river twice, for it's not the*
*same river and he's not the same man.*
— Heraclitus

A major theme of this book has been the affect that our relative hormonal make-up has on our perceptions of life and the world. Much evidence has been presented to exemplify how differently society organizes itself depending on the predominant hormonally-driven worldview. When a culture swings out of balance, it cycles into one extreme or another of a kind of manic-depression. As would be expected, our very capability to conceive how the physical world is composed is influenced by these cycles; and in this chapter, we will review how that reveals itself in yet another realm of science wars.

The branch of philosophy concerned with the nature of reality is called metaphysics. Heraclitus and Parmenides were pre-Socratic Greek philosophers who each reduced the world to one fundamental property; but their conceptions of reality were in fact opposite one another. Heraclitus focused on the constant state of change in the universe, represented by the energy of fire, while Parmenides believed that the universe is made of a static block of matter. Each of these great fathers of Greek philosophy identified with an aspect of nature they subjectively chose to focus on in their observations. The Western eye still struggles today with this conflict about the nature of reality. One side views the order set in the material world, while the other sees a universe of energy flow that is in a state of chaotic flux.

The origin of this conflict lies in the make-up of our brains, in the feminine versus masculine. In *Sexual Personae*, Camille Paglia argues that in Western culture, based on Greek philosophy, this dichotomy is represented by Apollo and Dionysus. The Apollonian stands for maleness, order, and structure in creation, while the Dionysian represents the female element of chaos, anarchy, and flux of motion in nature. Paglia attributes the biological impetus to this: "The quarrel between Apollo and Dionysus is the quarrel between the higher cortex and the older limbic and reptilian brains."[1] The systematizing male brain has elucidated the foundational taxidermies of the world we live in, as well as the mechanistic concepts that have ushered us from one degree of technological progress to the next. Be it the emergence of germ theory in medicine or the advances in physics that enabled mankind to fly to the moon, it was the attentive, objectifying male brain that perceived and conceived the patterns in the natural world that led to subsequent creative scientific advancement.

Feminists understand that the extreme male brain activated the rise of the modern scientific paradigm, but they have not been happy about it. Known as Newtonian classical mechanics, Newton's physical view of the universe is based on a male's objectifying perception of nature into distinct material objects. Objectification in social psychology is defined as relating to a person or an animal as a material object or a thing, and it is condemned by feminists, who see only the propensity for sexual objectification of women in this characteristic of the male brain. Research, however, shows "a clear schism" between how people actually perceive images of men and women in their brains. Both men and women are inclined to see women as their component parts, with the brain function of local processing; while men are seen as a complete body by the brain's global processing.[2] This would indicate that the "objectifying" per se is not the problem, rather it's the sexual attraction that men feel toward women. Feminism has conflated the components of perception and result. In *Sexual Solipsism: Philosophical Essays on Pornography and Objectification*, Rae Helen Langton states that objectification causes the reductionist view of a person into the body, or body parts, a model that feminist philosopher Martha Nussbaum depicts in more detail as she defines the male tendency for objectification:

1. **Instrumentality** – treating the person as a tool for another's purposes
2. **Denial of autonomy** – treating the person as lacking in autonomy or self-determination
3. **Inertness** – treating the person as lacking in agency or activity
4. **Fungibility** – treating the person as interchangeable with (other) objects
5. **Violability** – treating the person as lacking in boundary integrity and violable, "as something that it is permissible to break up, smash, break into"
6. **Ownership** – treating the person as though they can be owned, bought, or sold
7. **Denial of subjectivity** – treating the person as though there is no need for concern for their experiences or feelings[3]

Burgeoning feminist studies of the 1960s and 70s extrapolated the polemic of the objectifying "male gaze" from social relationships to other societal endeavors, including science. In this vein, adversarial feminist philosopher Sandra Harding called Isaac Newton's *Principia Mathematica* a "rape manual."[4] According to this radical feminist view, Francis Bacon stated that "knowledge is power," and man invented the scientific method in order to gain power over Mother Nature and rape her soul.

## Newton versus Descartes

To understand the evolution of modern science, and particularly the physical sciences, we will want to review how testosterone shapes our mind's perception of the nature of reality and how this has manifested itself throughout the ages. In *A God of Math & Order* Professor Peter Harrison makes the case that "the new science rode in on the shoulders of theological ideas," driven the Protestant Reformation:

> First, Christian thinkers applied God's sovereignty to the natural realm in a new way, asserting that nature was governed by God-designed mathematical laws. Then, concerned to protect that sovereignty against Aristotle's notion that natural entities possessed intrinsic drives, Christians began to strip nature of her divinity, positing instead mechanical processes.[5]

This positing of mechanical processes revealed another rendition of the Greek debate between order and flux, which in the Enlightenment took the form of Isaac Newton's ideas contrasted with those of Renee Descartes. Natural philosophers tended toward either the more abstract, Newtonian mathematical models or the more mechanistic models proffered by Descartes. The arena of debate was nowhere more pronounced than in considering how planets "hung" in the heavens: What is the nature of gravity? What comprises the space between planets? Cartesians considered the same phenomena as Newtonians, yet their preference was to understand causal, measurable, observable relationships between celestial bodies; whereas the Newtonians sought to describe the interactions mathematically. Through those lenses, the question of space was hypothesized by Newtonians as a void and Cartesians as matter filled.

For Descartes the existence of a substance was absolutely necessary for planets to move. The substance, or ether, in space "carries" or "moves" the entities in their orbits. The sun is to be seen as the center of a large vortex around which the ether swirls, moving the planets. Each planet functions at the center of numerous smaller vortices that their respective moons and comets move within. By contrast, Newton understood planetary motion in terms of elliptical trajectories maintained by centripetal force. In his writing *The Mathematical Principles of Natural Philosophy (Principia)* of 1687, Newton wrote of what could be quantitatively observed:

> The forces by which the primary planets are continually drawn off from rectilinear motions, and retained in their proper orbits, tend to the sun; and are inversely as the squares of the distances of the places of those planets from the sun's center.[6]

The potency of a quantitative explanation for Newton was sufficient. He was not compelled to hypothesize about causation, as he could hardly reach into outer space and collect a substance to show to the world. He was resolute that experimental philosophy be validated by measurements of phenomena, described with calculations, and then generalized by induction. The effects of gravity on celestial bodies did not require substance; rather, only empty space would allow for the unencumbered movement of planets in the solar system.[7]

Reading the metaphysical debates of the eighteenth and nineteenth centuries, it is readily apparent that accusing the other camp of "occult qualities" was the height of criticism. The term in context was somewhat different than it might be understood today. The notion was that something was hidden or concealed, while not necessarily magic or in some manner supernatural. It referred to phenomena that could not be directly seen or experienced, and therefore could not be empirically measured. Cartesians like the great German philosopher G.W. Leibniz considered gravity itself occult:

> The ancients and the moderns, who own that gravity is an occult quality, are in the right; if they mean by this that there is a certain mechanism unknown to them, whereby all bodies tend towards the center of the earth. But if they mean that the thing is performed without any

mechanism by a simple primitive quality, or by a law of God, who produces that effect without using any intelligible means, it is an unreasonable occult quality.[8]

Newton, on the other hand, and his school, regarded as occult those phenomena posited without evidence. While the causes of gravity or magnetism were not comprehended, their effects could be observed and measured, whereas Descartes' fluid or Leibniz's "monads" were merely inferred from the position and movements of objects.[9]

With the tools of modern neuroscience, we shall now utilize our knowledge of how sex hormones shape our perception of reality to analyze the forces that drove these science wars and the paradigm shifts in the course of scientific revolutions. It must be noted that the perceptions of the material universe from different perspectives actually complemented rather than contradicted each other. Like the yin and yang, or the female and male, in Chinese philosophy, both are considered equal forces of nature:

> The yin and yang depict how seemingly contrary forces may actually be complementary, interconnected, and interdependent in the natural world, and how they may give rise to each other as they interrelate to one another. Many tangible dualities (such as light and dark, fire and water, expanding and contracting) are thought of as physical manifestations of the duality symbolized by yin and yang.[10]

Descartes and Newton were decidedly minds of their times and their religions. Descartes' Catholicism was informed by an Aristotelian view of the universe, in which there is no empty space, but rather a continuous sphere of matter moving in perfect cycles. In contrast, the English Puritan Newton identified matter with objects, such as the earth rotating around the sun, but did not see any matter between those objects. Moreover, Descartes, who published the *Principles of Philosophy* in 1644, developed his work in the period of great turmoil and chaos, in which a negative trend in social mood brought the Thirty Years' War (1618-1648) that devastated Europe. Newton published his work in 1687, long after the war ended, with the Protestant worldview gaining over Catholicism and the Peace of Westphalia in 1648 creating the basis of national self-determination within secure borders of sovereign states in Europe.

A similar distinction can be seen in the difference between Catholic Renaissance art that reveled in pagan nature imagery and the Puritan iconoclasm, wherein religious images and icons were destroyed as impure idolatry against the incorporeal God of the Bible. The extreme masculine brain is interested in imposing order on the universe and refrains from submitting to the chaos of the material world as a form of submission to Mother Nature. The feminine brain is inclined more to beauty, because it is designed to attract the male eye for sexual pursuit. The Protestant eye views the world from a masculine perspective in philosophy, art, science, and politics.

According to the Testosterone Hypothesis, Descartes' work in this period of testosterone decline shaped his worldview to perceive the universe as in a state of a vortex motion. That model is far more chaotic than Newton's objectifying eye focusing on objects and matter and refraining to acknowledge or even postulate that there had to be some intervening material force or energetic particles behind gravitational attraction. Although both great Enlightenment thinkers had an Apollonian mind that sought to discover patterns of order and structure in creation, Newton's worldview is related to the more extreme male brain.

With Protestant Europe and Great Britain rising to become the leading force in the West, the Newtonian view gained acceptance, with the Western eye gravitating toward this extreme-male-brain view of reality. However, because the Protestant-led scientific establishment gradually abandoned the Aristotelian-Cartesian view of the universe as composed of ether and not mostly empty space, Descartes lost the great discovery of the vortex theory of matter.

## Light: The Origins of Particle-Wave Duality

True to his masculine, objectifying view of nature, Newton perceived light as particles in motion, rendering darkness as the lack of light. This metaphysical worldview is an essential part of English Enlightenment philosophy, which is based on the optimistic biblical conception of creation, identifying the Creator with light in Genesis 1. However, after 1776 and into the early nineteenth century, solar activity began to decline sharply in the Dalton minimum. Resultant of this decline was the German counter-Enlightenment revolution in philosophy that brought back the forces of darkness. One of the great German philosophers of the period was Johann Wolfgang Goethe (1749–1832), author of the influential novel *Faust,* in which man sells his soul to Satan in order to gain control of humanity and nature. While the English Enlightenment philosophers focused on the forces of creation, light, and the good associated with God as the foundation of the modern world, the German counter-Enlightenment philosophy came to command the Western consciousness, focused on a malevolent world of darkness, chaos, and destruction. This was depicted in the nihilistic, existentialist, postmodern philosophy of Friedrich Nietzsche, which is anti-existence and anti-life.

In 1810, Goethe published the book *Theory of Colors,* attacking the Newtonian view of light as particle and asserting an ambiguous position, building further on Kant's earlier attack against objectivity in the *Critique of Pure Reason* (1781), asserting the mind can never truly perceive reality. In the twentieth century, Ludwig Wittgenstein opined on Goethe's theory: "I believe that what Goethe was really seeking was not a physiological but a psychological theory of colours."[11] Wittgenstein seems to imply a subjectivity in Goethe's philosophy, one commensurate to the era of declining social mood in which Goethe was living. In stark opposition to his contemporary scientific consensus, Goethe elevated darkness to its own force. Rather than being perceived as the absence of light, darkness to Goethe was an opposing force that interacts with light and creates a shadow over it. Rudolf Steiner wrote on Goethe's view of darkness:

> Modern natural science sees darkness as a complete nothingness. According to this view, the light which streams into a dark space has no resistance from the darkness to overcome. Goethe pictures to himself that

light and darkness relate to each other like the north and south pole of a magnet. The darkness can weaken the light in its working power. Conversely, the light can limit the energy of the darkness. In both cases color arises.[12]

While Newton postulated light as particle, trusting his senses and logical reasoning, based on the experimental method, Goethe believed that we can only make inferences about the nature of light, moving from premises to conclusion with a lack of clear, sensory evidence. Furthermore, the English physicist Thomas Young made the famous double-slit experiment in 1801, from which he perceived light as a wave. Known as the wave theory of light, this model joined the assault on Newton's particle view of light. Young based his view on the idea of wave interference in observed physical phenomena, in which the interaction of waves in water, light, or other forms of matter propagates an effect. This happens when waves are superimposed and together form a wave of greater, similar, or lower amplitude as observed in ripple tanks.

The wave motion of particles, however, does not actually negate the particle theory of light any more than the wave form in which water moves does not negate the obvious fact that it is made of separate, physical water molecules. As will be discussed later, the wave form of matter is caused by its vortex motion, not because it is a mystical, supernatural phenomenon. Light moving in wave form is simply the phenomenon of particles moving not through empty space, but through an ether invisible to the human eye, unlike water molecules that we can observe. However, we have to remember that our sense of vision is very limited, and we can observe only a small fraction of the light spectrum (between 390 and 700 nanometers).

According to Goethe's contemporary Hegel, whose philosophy is a product of the same *zeitgeist*, history moves by a three-step process, thesis, antithesis, synthesis. This dialectic is a cycle of problem, reaction, and solution. It suggests that reality is internally contradicting, but the struggle leads to a solution at the higher level of unity between the contradictions. The history of physics seems to play out in this form: after the Newtonian paradigm of light as particle was rejected in the nineteenth century for light as wave, Einstein's 1905 discovery of the photoelectric effect led to a new synthesis of these approaches. Studying the emission of electrons when light shines on a material, Einstein reached the conclusion that light is both a particle and a wave, allegedly resolving the century old conflict. This spawned the twentieth-century contradictory belief system of the wave–particle duality, which states that light is at the same time both a wave and a particle. This negates logic and reality, as the light wave is actually the motion of a particle and does not have a spiritual existence of its own. Yet, as we persist in the process of the dialectic, one generation's synthesis becomes the next generation's thesis, open to investigation that leads to another antithesis, and so on.

## Matter and Energy

Material systems are made of vortex motion in which matter in motion has energy. As Einstein stated in his renowned mathematical formula, $E = MC^2$, energy is matter in motion. Matter is actually made of ever-smaller particles of matter on an ever-reduced scale that rotate around their center of mass. Like the planets rotate around the sun, electrons rotate around the nucleus of an atom, the fundamental, known unit of matter. In 1913, Niels Bohr developed this model of the atom, conceptualizing its form as similar to the solar system's, with the difference that the attraction mechanism is provided by electromagnetic forces rather than gravitational forces.

The masculine and the feminine brain have perceived different aspects of reality, while failing to connect the dots and integrate these parts to form the entire, true picture of matter in motion. In a unified schema, Descartes and Newton's theories are not contradictory, but each depicts different parts of the same dynamic physical system. The decline of testosterone since the year 2000 has incited a return to the pre-Newtonian, Cartesian scientific paradigm and will gradually change our view of the nature of our cosmos. In the groundbreaking book *Universal Cycle Theory: Neomechanics of the Hierarchically Infinite Universe* by Stephen J Puetz and Glenn Borchardt (2011), the authors describe vortex motion as the basis of universal matter and energy. Offering a kind of synthesis that depicts a "neomechanical" perspective, they assert that gravity comprises all the components historically attributed to it:

> Gravitation follows the inverse-square law, just as Newton said; it involves inertia, just as Einstein said; it involves pushing, just as Lesage said; it includes vortex motion, just as Descartes said; and it entails aether, just as many philosophers since the ancient Greeks said.[13]

While the authors' view of the universe as infinite is controversial, given that matter and existence can only be finite in objective reality, their revolutionary theory that the universe is made of vortices scaling in ever-greater hierarchical structures conforms to the fractal nature of reality according to chaos theory. Fractals are patterns that are self-recurring on different scales of time and space and are observed throughout the universe, from galaxies at the larger end, to our DNA structure, and even in smaller scales of matter.

In the early twentieth century, another period of declining testosterone levels, the Newtonian paradigm was replaced with Einstein's Relativity theory. Rejecting absolute time and space and replacing it with relative motion, the theory denies objectivity and reality in the physical universe for a more subjective perspective. Einstein's discovery that light curves during its travel and the subsequent experimental confirmation in 1919 by Sir Arthur Eddington made Einstein a celebrity. This precipitated the eventual triumph of general relativity over classical Newtonian physics. Even more problematic, however, as testosterone continued to collapse into the 1930s, Heisenberg's Uncertainty Principle became accepted as "scientific" dogma, altogether subverting reality into an ultra-subjectivist philosophy by claiming that a particle can be in multiple locations at one time, only collapsing to a single state when being perceived by a human observer. Heisenberg's ideas pushed Western science to embrace the concepts of Eastern mysticism with the rise of the uncertainty principle as the foundational principle of reality. This was confirmed in interviews with Heisenberg by physicist Fritjof

Capra, who wrote the *The Tao of Physics: An Exploration of the Parallels between Modern Physics and Eastern Mysticism* (1975).[14] Heisenberg accounted for the influence Eastern philosophy had on him as he travelled and lectured in India. Not only the declining testosterone of the era, but also Eastern mindset overall favors the more feminine brain type that is anti-objectivity and tends to deny objective reality

However, in reality, our universe is a dynamic fractal system, and our solar system itself is a vortex travelling through our spiral Milky Way galaxy, orbiting around the center at a speed of approximately seventy-two-thousand kilometers per hour, while the planets orbit the sun.[15] This paradigm shift is gradually becoming popularized through the stunning video animation modeling of the Helical Model of the universe done by DJ Sadhu.[16]

### The Electric Universe

A foundational tenet of this book is that declining solar radiation reduces the happy chemicals in our brains, causing turmoil in our natural and social environment. Commensurate with these physiological changes, our cosmology tends to change in the direction of a more chaotic view of the universe, such as the chaos theory popularized during the 1970s. Newton's orderly universe, controlled by the masculine personality-type Creator gives way to a chaotic flux of energy and matter such as with plasma and ionized gases. This fourth state of matter that exists in space is produced by extremely strong electromagnetic fields that separate positive ions from negative electric charge. This plasma state of matter is considered the most abundant form in the universe.

In a period of negative social mood, during the 1740s, Benjamin Franklin's mind turned to understanding the vicissitudes of polar lights in "motions of the luminous matter of the aurora." He wrote about his speculations in *Aurora Borealis, Suppositions and Conjectures towards Forming an Hypothesis for its Explanation,* to explain the phenomenon observed in the northern regions.[17] Again in the early twentieth century, Kristian Birkeland (1867–1917), a Norwegian scientist, published a book in 1908, *The Norwegian Aurora Polaris Expedition 1902–1903*, stating that the polar lights are caused by electric field lines formed by the sun–earth electromagnetic connection through the polar regions. These electric currents are now called Birkeland currents in his name, but his theory was not accepted until in 1967, when a probe was sent to space and discovered the aurora, using a magnetometer above the earth's ionosphere.

The Swedish plasma physicist Hannes Alfven received a Noble Prize in physics in 1970 for his work on magnetohydrodynamics, including his theories describing the earth's magnetic field, magnetic storms, and plasma dynamics across our Milky Way galaxy. Although many of his theories became accepted into the 1980s, in a 1986 paper, *Double Layers and Circuits in Astrophysics,* he stated that his electromagnetic view of the universe is much neglected in academia:

> A study of how a number of the most used textbooks in astrophysics treat important concepts such as double layers, critical velocity, pinch effects, and circuits is made. It is found that students using these textbooks remain essentially ignorant of even the existence of these concepts, despite the fact that some of them have been well known for half a century (e.g., double layers, Langmuir, 1929; pinch effect, Bennet, 1934).[18]

Since 2000, as social mood is declining again, there is some rising activity among the "electric universe" community, but the gravitational model of the cosmos, fathered by Isaac Newton and transformed by Einstein's theory of relativity, still reigns supreme in the mind of Western academic science and popular culture. In the super organismic view of our social structure, archetypal change in our collective unconscious views of the nature of existence as a process that is controlled and regulated by our hormonal cycles over long periods of time. However, as our social mood deteriorates, the chaotic, electric cosmology, more entwined with the feminine, Eastern philosophies and focused on energy flow and the unity of the cosmos will rise again.

# Sex Wars

By integrating the understanding of how sex hormones shape our mind and perception of reality with the history and philosophy of Western science, we can now reach a more comprehensive, objective level of scientific discovery, while making the necessary corrections for our psychological biases and group think. The yin and yang, the masculine and feminine, active and passive, are forces that shape our views of the universe. By studying our own brains as an integral part of the mechanical universe, a mechanistic philosophy of nature, which both Descartes and Newton championed, we can ultimately discover the secrets of the universe and advance objective and empirical science rather than surrender our mind in submission to mysticism and collective delusions.

## Phase Transitions in Water and the Evolution of Life

Life as we know it on earth is made of water in the liquid state. We know of three primary phases of matter: solid, liquid, and gas. These states of matter are presented in the philosophical argument between the two extremes, on the one hand, Parmenides, who viewed the universe as static, such as the solid state of matter, and on the other, Heraclitus, who viewed nature as completely fluid, such as the gaseous state of matter. Life exists in the middle phase of liquid, requiring the essential property of fluidity that enables a degree of stability, but not too static to repress the dynamism to evolve, to change through time. At the extremes, solid matter is too static for life, and the gaseous state is too fluid to allow for any stable structure to form.

Earth is the only planet we know in our solar system that can support life. Its proximate distance, about 150 million kilometers, from the sun is a significant factor in making it inhabitable. Earth's atmosphere also fills a vital role in regulating its temperate and hospitable climate by providing the gasses required to protect us from excessive radiation and heat from the sun and also trapping heat in the atmosphere to allow for a stable temperature range. For instance, the next closest planet to the sun, Venus, has much too hot an atmosphere, as temperatures there reach more than 750 F (400 C). The next farthest planet from Earth, Mars, which lacks atmosphere, is too cold, as its average temperature is minus 80 F (minus 60 C). This explains why a stable atmosphere on earth was required for the growth and flourishing of life here. The atmospheric gasses retain heat from our sun, with an average global temperature of 61 F (16 C). The earth's average temperature is above 32 F (0 C), the freezing point, and much less than 216 F (100 C), the zero point of boiling into vapor, and hence allows for water to remain in a liquid state, as required for life.

The evolution of life is a product of the forces of energy working on material objects, the living beings that evolve through time. DNA is the genetic code of life, the genotype supplies the instructions for the living cell on how to acquire, use, and adapt to the energy states supported by its environment. In complex life, both animals and plants, we observe the changes driven by solar radiation cycles, such as the daily, monthly, and seasonal earthly cycles in temperatures.

Biological life is designed to constantly evolve in order to adapt to environmental change. Life is the process of sustaining and furthering itself; it is matter in motion, responding to energy flows. The living organism is an object with a form and identity that is open to change. Hence, is it both static, having a coherent structure at a given point in time, and dynamic in its fluidity, the ability to change form through time. The mechanisms of change are both genetic, such as genetic mutations and recombination of genes in sexual reproduction, and epigenetic, the turning off and on of certain genes to alter phenotypic behavior patterns, usually driven by hormonal signals in complex multicellular life.

DNA codes for the physical structure of the phenotype. In 1944 the physicist Erwin Schrödinger wrote *What Is Life? The Physical Aspect of the Living Cell.* In the book, he focused on chemistry and physical processes of life and introduced the idea that genetic information is contained in an "aperiodic crystal" through the configuration of its chemical bonds. His ideas later inspired Watson and Crick in their discovery of the structure of the DNA molecule in 1953. Just as the letters of the DNA code signify the production of life's complex proteins on the biological level, on the chemical level, the periodic table of elements comprises the list of various atoms that can bond together to form complex molecules. The atom, in turn, is formed on the physical level of complexity by the bonding of the proton and the neutron to form the atomic nucleus, attracting the electrons that orbit around it.

According to the photoelectric effect, photons of light, particles that are radiation of electromagnetic energy waves, hit the atom and change the energy state of the electrons in their orbit around the nucleus. Even the fundamental particles, such as protons, neutrons, and electrons have spin and may be composed of smaller particles orbiting in their microcosms. The ether, according to ancient and up to nineteenth-century science, is the invisible matter in space.

Hence, to borrow a metaphor from biochemist Albert Szent-Györgyi, matter is energy dancing to the tune of light waves. At the most fundamental level, matter is itself made of particles in motion. The energy level of a particle is determined by its mass and velocity of motion, which we define as its ability to do work or to change the state of another particle by physically interacting with it.

It can therefore be said that our conceptual separation of matter and energy is artificial and may often lead to error in understanding the nature of the universe. In reality, there is only matter in motion that interacts with other particles in motion. Energy is a useful concept for calculating the ability of matter to do work, but it should not be confused as something non-material. Likewise, the concepts of time and space are used to measure the period and distance, accordingly, of particle motion, but are often confused as having an existence of their own, such as in Einstein's warped space-time, as if there is a fourth dimension to reality that makes it relativistic by gravitational fields. However, Einstein himself noted that his model that replaced these theories could itself be thought of as an ether, because it implied that the empty space between objects had its own physical properties.[19] Aristotle described the spheres as held together by ether in a perfect circular motion. In the eighteenth century, the French scientist Johann Bernoulli theorized that all space is composed of ether in "excessively small whirlpools" that allow in their elasticity for vibrational transmissions of light.[20] In twentieth-century physics, confusion has caused the rise of peculiar belief systems to explain observed phenomena in academic science, such as the postulation of dark energy and dark matter to explain the gravitational force of galaxies, and the wave–particle duality that suggests light is both a particle and a non-particle wave at the same time.

The misperceptions in physics arise from flaws in the definition of concepts, grounded in a false philosophical foundation of science. Such confusion in man's thinking is exacerbated by how sex hormones drive our perceptions of reality and shape our cultural views of existence. The fact that scientists have often disregard the vortex motion of particles in the universe, which orbit around a center of attraction, the nucleus, is evidence of this. Moreover, the nature of the universe is fractal, meaning the vortex motion is self-repeating on different scales, from the vortex of electron orbits around the nucleus of the atom to the planets orbiting around the sun, to our solar system orbiting around the spiral Milky Way galaxy, to the galaxies themselves orbiting around a greater concentration of galaxies in the universe. In other words, fractals exist from microcosms to macrocosms.

A premium example of a scientist thinking independently of the herd is Gerald H. Pollack. A professor of bioengineering at the University of Washington, Pollack wrote *The Fourth Phase of Water: Beyond Solid, Liquid, and Vapor*, in which he presents his revolutionary ideas on how energy drives the structure and phase transitions of water and, by extension, our biology and cell life, which is mostly made of water molecules.[21]

In his book's preface, Pollack criticizes the postmodern scientific culture that is fundamentally at odds with logic and reality. He resists "the notion that modern science must lie beyond human comprehension."[22] He rejects principles that cannot be explained by observation, seeking more straightforward explanations, even at the peril of overturning "towering scientific giants."[23] In Pollack's analysis of water, per the title of his book, he introduces the ingenious concept of water's "social behavior."[24] As a constellation of separate water molecules, it functions in ways akin to the complex behavior of starlings, for example. It ebbs and flows and swirls and undulates in ways that a single molecule does not. Pollack rightly critiques the scientific deductive method that has sought to extrapolate from one water molecule's interactions with nearby molecules to larger bodies of water. His method is to consider the peculiar "crowd" behavior of water as its own dynamic, rather than scoping out from individual molecules.

Furthermore, Pollack attributes to water an energy storing capacity that has potentially profound implications for how we might harvest energy from the water surrounding us—perhaps even within our own bodies. Water transitions between phases, as noted above: solid, liquid-crystalline, liquid, and vapor, and within the vapor phase, a "ready source of electrons that could drive any of numerous biological reactions."[25] He explains the phenomenon like this:

> Water's fourth phase stores energy in two modes: order and charge separation. Order constitutes configurational potential energy, deliverable as the order gives way to disorder. For the working cell, this order-to-disorder transition constitutes a central energy delivery mechanism. Charge separation, the second mode, entails electrons carrying the EZ's usual negative charge, while hydronium ions bear the corresponding positive charge. Those separated charges resemble a battery — a local repository of potential energy.[26]

In addition he states that the sun's electromagnetic energy builds potential energy in water:

> To suggest that incident solar energy may build energy in our bodies may seem a stretch, but cells do grow faster with warmth, i.e., when exposed to infrared energy (light). Since light builds energy in water, and we are mostly water, it seems plausible that we might harvest energy from the environment. Multiple light-harvesting mechanisms can be envisioned throughout biology.[27]

# 9 • THE PHYSICAL UNIVERSE

Professor Pollack seems, as a high-testosterone science personality, more interested in groundbreaking scientific principles than serving the feminine culture associated with the hive mind and group think. As he describes in his metaphor of the scientific tree of knowledge, the major branches of science were developed by revolutionaries of independent mind who were concerned with the fundamental understanding of natural principles. This is typical of the systematizing extreme-male brain. However, as testosterone levels decline, we are now growing only the twigs on the tree of human knowledge, as opposed to growing the major branches that hold the tree's structure by major figures who laid the foundational principles of modern science, primarily Newton and Darwin. When testosterone collapses, our civilization will fall with it into darkness, and the tree of knowledge we created will wither out and die with it. However, the DNA, or information contained within its seeds, our books that hold our knowledge, will continue to lie dormant until another phase of growth resumes, and this knowledge will be transmitted through manuscripts to enable the growth of another human civilization.

## The Epic of Evolution

Incorporating the energy-centric perspective of thinkers like Pollack, there is a new, interdisciplinary narrative of cosmic, biological, social, and cultural evolution being formulated. The idea of an "Epic of Evolution" is a product of the fruitful mind of E. O. Wilson, who used the term "evolutionary epic" in 1978. Prominent scholar, Eric Chaisson, an American astrophysicist, has subsequently developed the interdisciplinary science of cosmic evolution. The historian David Christian initiated the emerging academic discipline called Big History, studying the evolution of the universe from its creation in the big bang to the present, exploring common themes and patterns.

Cosmic evolution, promoted by natural scientists, is more concerned with physical, quantitative analyses of metrics, such as the energy flow that drives the evolution of complexity in the universe.

According to Chaisson, "A wealth of observational data supports the hypothesis that increasingly complex systems evolve unceasingly, uncaringly, and unpredictably from big bang to humankind."[28] Big History, closely overlapping in many aspects with cosmic evolution, is studied by social scientists, and hence is more focused in its scope on human civilization and culture. Like Pollack, Chaisson has a bias toward simplicity—"a remarkable simplicity that underlies the emergence and growth of complexity for a wide spectrum of known and diverse systems."[29] In "The Natural Science Underlying Big History," he explains how the rate of energy flow, specifically *energy rate density*, determines the level of evolutionary complexity in the studied emerging systems. To unify the measurement of energy flow in different fields of the natural sciences into a single, interdisciplinary method, the term energy rate density was coined, being applied to the study of cosmic evolution in general. Energy rate density defines the quantity of energy flowing per unit mass in a unit of time, proposed as a quantity to assess the complexity of the system. This is a general term, but is equivalent to more specialized terminology used in various scientific disciplines. For instance, in astronomy it is the luminosity-to-mass ratio, in physics the power density ratio, in biology the specific metabolic rate, in geology the radiant flux per unit mass, and finally in engineering the power-to-weight ratio.[30]

Chaisson characterizes the correlation between rate density and evolutionary success by stating, "Operationally, those systems capable of utilizing optimum amounts of energy tend to survive, and those that cannot are nonrandomly eliminated."[31] Figure 14 illustrates the function of energy rate density across time, showing the exponential increase in complexity. Indicated on the graph, for example, is how cultural evolution moves faster than biological evolution (note steepest line on right side). The dashed lines in the shaded area represent such evolutionary events as the evolution of stars from birth through death, of trees from grasses or the human progression from hunter-gatherers to agriculturalists and onward.

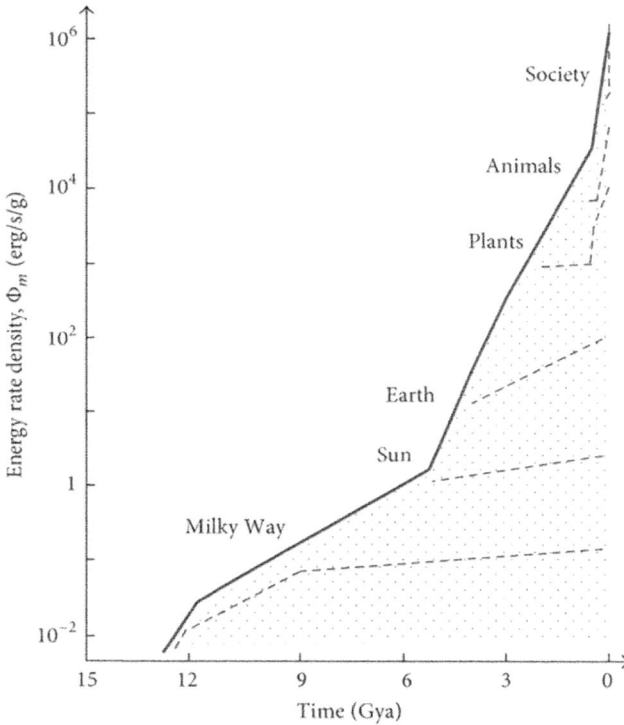

Figure 14: Energy Rate Density and Evolution (Copyright © 2014 Eric J. Chaisson) [32]

One example of an analytical tool that has aided social scientists, especially, in conceptualizing energy flow within human cultural evolution, is the Kardashev scale. This scale was created in 1964 by the Soviet astronomer Nikolai Kardashev, and it measures the technological advancement of a society by measuring the energy used for communication. It claims three types of civilizations:

A Type I civilization – also called planetary civilization – can use and store energy which reaches its planet from the neighboring star.

A Type II civilization can harness the total energy of its planet's parent star (the most popular hypothetical concept being the Dyson sphere—a device which would encompass the entire star and transfer its energy to the planet(s)).

A Type III civilization can control energy on the scale of its entire host galaxy.[33]

## The Problem with Entropy: The Second Law of Thermodynamics

To understand the cultural mindset that compels scientists to see through a lens of chaos, we turn to the condition of entropy—the degree of disorder in a system. Specifically, entropy became codified in the mid-nineteenth century through the work of Lord Kelvin (best known for determining the value of absolute zero) in the second law of thermodynamics. Simply defined in *Wikipedia*, the law means:

> The total entropy of an isolated system always increases over time, or remains constant in ideal cases where the system is in a steady state or undergoing a reversible process. The increase in entropy accounts for the irreversibility of natural processes, and the asymmetry between future and past.[34]

This is to say that in a closed system, disorder always increase unless outside energy reaches into the system to increase order. The increase in disorder in the universe throughout time is thus deemed as inevitable, because the universe is all that exists, and no outside energy can enter into it to defy the forces of chaos. In 1851, referring to energy in thermodynamics, Lord Kelvin stated, "It is impossible, by means of inanimate material agency, to derive mechanical effect from any portion of matter by cooling it below the temperature of the coldest of the surrounding objects."[35] The German Rudolf Clausius, examining the relationship of heat transfer and work a century later, stated his formulation in 1954, "Heat can never pass from a colder to a warmer body without some other change, connected therewith, occurring at the same time."[36] It was Ludwig Boltzmann (1844–1906), the developer of statistical mechanics, who related the second law of thermodynamics to a law of disorder as a result of his atomic view that mechanical particles hit in random collision until they reach the state of maximum disorder, or entropy.

The depressed social mood and despair of the 1850s has been referred to repeatedly in this book. It was, of course, the period in which Karl Marx formulated the doctrine of communism in class struggle and Darwin formulated the evolutionary concept of the struggle for the survival of the fittest. The second law of thermodynamics, likewise, was conceived in this decade, and has been portrayed by one author as "perhaps the most pessimistic and amoral formulation in all human thought,"[37] given that it dooms the universe to ending in utter chaos and entropy.

In contrast to this pessimism, however, and in reality, according to Puetz and Borchardt's Universal Cycle Theory, there is no empty space. The universe, rather, is composed of hierarchical vortex structures. These structures orbit in different scales of size, in states of accretion and excretion of matter according to energy flows in each microcosm in relation to the macrocosm of the fractal system. In periods when energy flows into the system and it accretes new particles, its energy level and order grows; while in periods when energy flows out of the system as it excretes particles, the degree of entropy increases and its order and energy levels are reduced. This is a periodic, cyclical phenomenon that recurs throughout nature, forming the fractal geometry of nature as a spiral design. These are the phase transitions of nature between orderly matter and chaotic flux that are driven by energy flows, similar to the form matter takes as shaped by the energy levels of its electron orbiting around its center. This is demonstrated in Einstein's photoelectric effect, in which the state of matter, the atom, changes when energy levels change.[38]

# Sex Wars

In the Newtonian paradigm of empty space around the star and planets of a solar system, the thinking is that matter always travels from areas of concentration to areas of less concentration as the system tends toward equilibrium and entropy. Entropy seems to be a fatal lack of order in the system as it degenerates from order, because different elements move apart from each other and apparently move toward a kind of perfect, internal disorder. They are dispersed chaotically, mixed up with no coherent structure. However, in actuality, there are ether particles with velocity and mass, and these particles are in motion and bond with other particles to form vortex structures that can be identified as objects on a larger scale. A prime example of this is the planets that orbit around the sun to form the solar system, a composite object larger than its components.

## Testosterone - The Energy Regulation Hormone

As we have learned throughout this chapter, life is a process that requires the constant flow of energy to sustain and further its existence. Understanding how the living organism, a material object, utilizes energy flows in order to live and reproduce is crucial to studying the evolution of life, how sexual reproductive strategies drive the sexual organization of complex social systems, and the origin of the reproductive life cycle with its phase transitions of life, sex, and death.

Primary hormones that regulate the body's energy levels are serotonin, testosterone, and thyroid that influences the metabolic rate and protein synthesis. An inefficient level of thyroid production, hypothyroidism, is also associated with low testosterone levels, the hormone that builds strength in bone and muscle growth. Both hormone levels are regulated by the hypothalamus, the master hormone regulator in our brain. While hypothyroidism is usually associated with women, men who suffer from this deficiency are found to have a condition called hypogonadism that results in reduced testosterone concentrations.[39]

# 9 • THE PHYSICAL UNIVERSE

The connection between our metabolic rate, governed by thyroid in particular, and the larger effect of energy availability is significant. The basal metabolic rate, measured in a resting state, is the energy used by an organism per unit of time. Some seventy percent of that energy is utilized by fundamental life processes, with another twenty percent used in physical activity, and a full ten percent required to digest food. This explains why testosterone levels are synchronized by levels of solar energy: our growth cycles are tuned to occur when energy availability is higher and there is more food available to fuel growth and anabolic processes. Hence, the brain is programmed to release happy chemicals and growth hormones when it senses rising levels of light, promoting growth and reproductive activity; conversely, the brain reduces these growth factors when light is reduced and darkness takes over. The phase transitions between periods of positive and negative social mood are modulated by these hormonal cycles. Positive social mood, encouraged by rising solar energy, brings a spirit of individualism, liberty, and capitalist pro-growth social policies, while negative social mood leads to more collectivism and socialist, anti-growth as well as constrained reproductive activity.

In the collectivistic phase, fascist and communist regimes rise to power, because our brains feel threatened by the limited availability of energy resources, and individuals organize into social groups that aggressively fight with each other for dominance over territorial, nutritional, and reproductive resources. This is the Darwinian contest for the survival of the fittest in an environment of constrained resources. It leads to the pessimistic socialist ideology of group conflict, seeing the world as a zero-sum game in which the gain of one group comes at the expense of another. In this phase, a lack of energy means there is no possibility to grow more resources, and there is an increasing reduction in productivity.

In conclusion, comprehending the causal links in the evolution of life, how energy flows determine the phase transitions of our social organization, is vital to achieve proper understanding of the relations between our philosophy of ideas, the biology of hormonal cycles, and the physics of energy and matter.

# Sex Wars

# 10

## TRANSCENDING DETERMINISM

*I, however, believe that there is at least one philosophical
problem in which all thinking men are interested ... the
problem of understanding the world – including
ourselves, and our knowledge, as part of the world.*[1]
— Karl Popper

*We cannot understand the world and ourselves
separately from it – human beings and the knowledge
they have – are intrinsically part of the world they are
trying to understand. This embeddedness of human
beings in the system they are seeking to understand, the
limits to their knowledge of that system, the impact of
their actions on the system's path, and the self-referential
circularity this inherently creates all lie at the heart of
George Soros's concept of reflexivity.*[2]
— Eric D. Beinhocker

Throughout this book we have discovered how sex hormones shape our lives, both as individuals and as social organisms that create extensive social networks to become part of a greater superorganism. The apparent contradiction between our social nature to conform and our will for self-interest, independence, and reason has been the primary force in the historical human struggle and the progress of our civilization. As an ultra-social species, mankind has created many forms of religious traditions and belief systems that determine our social organization, shape our collective consciousness, and guide individuals to become part of a greater whole. This concluding chapter will offer a review of many ideas from the previous chapters, as well as some thoughts for how we, as a species and as a culture, might proceed.

Since 2000, with the decline in social mood, the rationalist, pro-science, and individualist approach elevated during the past fifty years by great Western philosophers and scientists, such as Ayn Rand and Richard Dawkins, has given way to a more group-think view of society as an interconnected social network that shapes our individual minds. This has been a return to Freudian concepts from the early twentieth century regarding sex and civilization as well as Jung's psychology of collective consciousness that influenced the cultural Marxist philosophy of such influential thought leaders as Herbert Marcuse in his 1974 book, *Eros and Civilization: A Philosophical Inquiry into Freud.* Furthermore, evolutionary biologist David Sloan Wilson, similarly to psychologist Jonathan Haidt, asserts that humanity has evolved to be religious. He writes in *Darwin's Cathedral: Evolution, Religion, and the Nature of Society* that if we understand society as an organism, we can "think of morality and religion as biologically and culturally evolved adaptations that enable human groups to function as single units rather than mere collections of individuals."[3]

As the chief proponent in academia of the atheist, evolutionary-based school, Dawkins viciously attacks the Western, Judeo-Christian heritage in his 2006 best-selling book, *The God Delusion*, as anachronistic, primitive, anti-reason, and anti-science. However by 2016, even the liberal Dawkins came to realize that the Judeo-Christian faith in the masculine God, the Father of the family of mankind, is the best defense against the feminized West falling in surrender to the Islamic cult of submission. As reported in *Christianity Today*:

> Dawkins noted that Christianity, unlike Islam, does not make use of violent methods to fulfill its teachings. "There are no Christians, as far as I know, blowing up buildings. I am not aware of any Christian suicide bombers. I am not aware of any major Christian

denomination that believes the penalty for apostasy is death," he said. He admitted that he has "mixed feelings" concerning the decline of Christianity, because this faith-based group might just be "a bulwark against something worse."[4]

Dawkins is even now considered too masculine and pro-Western by the millennial generation that is de facto in control of academia. He was disinvited in 2017 from speaking on campuses such as the leftist UCal Berkley because certain comments about Islam "hurt people." The *New York Times* reported that the liberal public radio station as well as the university that were hosting his talk did not feel obligated to air his views in light of broad distress over his "Islamophobia."[5]

Notwithstanding Dawkins' fall from the graces of the politically-correct progressives, his anti-religious campaign has been termed a "rationalist delusion" by Jonathan Haidt—a "mass delusion," even, that plagues the elite institutions of Western culture.[6] This mass delusion of Dawkins and others is an assault on the masculine archetypical figure of God the Father. Moreover, it is of course ironic: the father of modern science, Isaac Newton, like all other major scientific figures of his time, was a devoutly religious Christian, who based his entire worldview on the Bible.

However, this attack on Christianity is not new. For example, it mirrors of the feminine "Cult of Reason" that led to the Reign of Terror and the guillotines during the period of the French Revolution, followed by the Napoleonic Wars for empire. The Cult of Reason was France's first established state-sponsored atheist religion, intended as a replacement for Roman Catholicism during the French Revolution. An excerpt from my first book, *The Objective Bible*, describes the movement:

> The new state declared a *Festival of Reason* in 1793 to celebrate the new religion, whereas, in actuality, the Age of Reason had already peaked a century before, and Western philosophy had been degenerating into pantheistic, mystical whim-worship of *un*reason. The

high altar at Notre Dame Cathedral was dismantled, and an altar to Liberty was installed with the inscription "To Philosophy" carved in stone over the cathedral's doors. The proceedings concluded with the appearance of a Goddess of Reason who, supposedly to avoid idolatry, was portrayed by a living woman. The underlying theme of the new faith was genuinely declared by Anacharsis Clootz, "one God only, Le Peuple"[7]

The pagan socialist cult ushered in a state of complete chaos, brutality, and anarchy. Thus, the people, whose collective will became god, asked for a strong man to unite them as an ideal society. In answer to their prayers and to terminate the chaos, Napoleon Bonaparte, who was heralded as the anointed savior of the people, consolidated power and formed a militaristic state while declaring himself Emperor. Napoleon embarked on a series of megalomaniacal war campaigns, devastating France and Europe while millions were killed in battle, ending in further disaster and bloodbath.[8]

It should be reiterated, certainly, that not only the extreme male brain is capable of reason. It was actually a drop in testosterone that led to the acceptance of Darwin's theory of evolution. Therein, Mother Nature battles against the authority of the masculine Creator in an eternal struggle between the creative forces of individual entities and the forces of their destruction. Out of this battle, a new whole emerges. However, if testosterone, along with other happy chemicals in our brains, becomes depressed below a certain threshold, a total assault on reason ensues. This is what was seen in the French Revolution and what can now be seen in the cultural revolution of the new millennium.

## The Philosophy of Evolving Complex Systems

If we examine our society from a viewpoint of complex biological systems, then our evolved psychological behaviors are adaptations to changing living conditions. Under this evolutionary worldview, we are simply biological structures programmed to evolve according to natural cycles. Evolution is not concerned with the wellbeing of individuals. Rather, as Hegel described, nature is a synthesis of opposing forces, a process of consuming the existence, identity, and independence of organisms to create a greater whole. Libertarian author Jeffry Tucker writes that Hegel's collectivist view of history entailed a group struggle for social domination that eventually led to modern socialism. Tucker explains that Hegel "abstract[ed] from human experience to posit warring life forces operating beyond anyone's control to shape history."[9] As discussed in the previous chapter, Hegel favored the philosophy of Heraclitus that views nature as an eternal process of change rather than one of static existence. This is supported by scientific evidence that the vast majority—in fact ninety-nine percent—of species went extinct through the Darwinian process of natural selection.[10]

Moreover, it was a dark social mood that led Darwin to adopt this evolutionary ideology that guided our social development. In his 1871 writing, *The Descent of Man*, he discloses some of his more controversial views, including the suggestion that society tends to coddle the weak by social welfare and charity, thereby "propagating their kind." Darwin's concern seemed to be that mankind was undermining its own genetic imperative for the most fit to prevail. Not surprisingly, ideas like this fueled the coming generations of eugenicists who "immediately began to plot demographic planning schemes to avoid a terrifying biological slide to universal human degeneracy."[11]

The feminine, fluid, holistic view promoted by these nineteenth-century thinkers, among others, perceives nature in flux and change, in which there is no black and white, merely grey areas in between. In contrast, the high-testosterone masculine mind has a self-centered view of the living organism, pursuing life by promoting values and a certain moral clarity that distinguishes between good and evil. Yet, as the evolutionary biologist Lynn Margulis discovered in her theory of Symbiogenesis, Mother Nature is a devouring mother who consumes her own children in order to adapt to changing environmental conditions. This process of experimentation leads to more complexity through the symbiosis of different organisms to create a greater whole, or superorganism. The problem is that this evolutionary process requires much sacrifice in order for only a few viable new adaptations or more complex life forms to emerge. Here it becomes clear what the origins of the sex wars were: the conflict between the Judeo-Christian, biblical worldview of the family of mankind under God, the Father of creation, commanding clear moral laws for his children that create an ethical order for society and the pagan, Gaian concept of an amoral, evolving system that devours its own creation in everlasting cycles of creative destruction.[viii]

Decline in testosterone levels explains why the science of complex systems has only come about in the last few decades, replacing the former Newtonian system of an orderly universe, guided by Divine Providence. As the previous order of our biological system falls apart, due to changing hormonal cycles, we view morality in terms of a grey area between black and white, or good and evil. Again, this explains the wild popularity of *Fifty Shades of Grey* in the feminist culture. As solar activity that generates light and thereby testosterone continued to wan, the next book in the series became *Fifty Shades Darker*, which suggests an even darker social mood.

In this culture of transition that produces much hormonal and genetic ambiguity, we have the transracial, transgender, trans-species and trans-humanism movements. Each of these depicts a state in which nature experiments with life to create new adaptations that will be able to survive and reproduce in the new environment. As noted, eugenics was one strategy by which people tried to control the evolutionary process of our species. Yet another was the Soviet experiment of so-called "positive eugenics" attempting to produce an ape-man. Crossing the genes of humans and chimpanzees, the plan

---

[viii] A comprehensive presentation of these ideas are available in my book *The Objective Bible: Western Civilization's Struggle for Philosophic Liberation from a Herd-mentality and Pagan Mysticism* (2014). See especially Part 1.

was "to speed up the spread of desirable traits–a willingness to live and work communally, for instance–and to get rid of 'primitive' traits such as competitiveness, greed and the desire to own property."[12] The Soviet endeavor to "realise their dream of a socialist utopia" might now seem almost regressive or possibly naïve, given that chimps themselves are aggressively competitive, yet they lack any rational capacity to overcome their instincts for dominance. Following a similar inclination in a different direction, the new millennium finds the trans-humanist movement seeking to merge humans with machine intelligence and robots. The AI expert David Levy expands on the fluidity of sexual identity and attraction in his book *Love and Sex with Robots: The Evolution of Human-Robot Relationships* (2008). Why should technological entities not be the objects of our companionship and even desire? Our increasing comfort with and reliance on technology could naturally evolve into amorous relationships with robots, especially as "sexual technology becomes increasingly sophisticated."[13]

The study of complex systems that has been developed in recent decades as testosterone levels plummet has inclined us to view our biological nature as a complex, adaptive, living system with recurring periodic cycles of growth and decay. Symbiogenesis creates new complex organisms from the merging of different forms of life. Inasmuch as this worldview dominates academia, Jonathan Haidt is trying to counter the homogeneity of belief and lack of diversity through his recent initiative of the Heterodox Academy.[14] However, history teaches us that our social superorganism evolves different perspectives on the world as a result of changing hormone levels that shape our minds' perceptions of life. In the course of our cultural evolution we create new syntheses of old belief systems. A prime example of this was the Judeo-Christian synthesis that came out of the period of ancient Israel, with its peak-testosterone culture that identified exclusively with the archetype of God the Father, to the more subtle, inclusive nature of Christianity. Christian theology believes in the Trinity, including the father principle in the Creator of mankind, the mother principle as the Holy Spirit that unites humanity, and the Son Jesus Christ who represent the united body of the Church.

Christian theology depicts the struggle of human life as spiritual warfare: the forces of Creator God versus Satan, light versus darkness, good versus evil, life versus death, and creation versus destruction. Although these religious archetypes are considered mystical forms of a scientific perspective, as we have discovered throughout this work, they actually depict the underlying psychological factors that stimulate our cultural evolution and the historical transformations in mankind's affairs. The physical forces of light and energy drive the evolution of matter through structural phase transitions, and likewise, our hormonal cycles are driven by energy flows of light, causing our master hormone regulator, the hypothalamus, to release more growth hormones and bring a period of increased growth and reproductive activity. From a psychological point of view, we feel happier. During periods of a rising wave of positive social mood, even our philosophy of life becomes optimistic, perceiving the universe as benevolent and moral. Accordingly, the Enlightenment—literally from "light"—is often contrasted with the Dark Ages; and as we have seen in the sunspot and temperature charts of chapter 7, these periodical transitions in social mood were actually driven by solar cycles.

Hence, our religious, social, and philosophical systems are synchronized with global energy flows that drive the biological life cycles on earth, tuned in to periodic changes in plant growth that supply energy resources from the animal food chain system. Sex hormones drive our social organization, whether toward greater growth and reproduction or toward hibernation, like polar bears in the winter, when energy and food resources are scarce. The purpose of this book has been to offer a perspective of the mind–body connection that integrates the natural sciences with the humanities to explain how our natural cycles are associated with physical forces that propel biological and even historical periods. By this investigation, a causal relationship between biological and cultural evolution has been revealed.

# 10 • TRANSCENDING DETERMINISM

In this book we have traveled a great journey through the ascent and fall of man, studying different aspects of human existence, from the physical and biological to psychological factors that shape our mind and philosophy of life. We have seen that man is a very social animal that forms social groups and establishes dominance hierarchies in respect to other groups; and our dominance posture in turn affects how we, as members of humanity, perceive our place in the natural order of the universe.

A people can ascend in the global pecking order of nations, such as the ancient Israelites did in the period of the Hebrew Bible, believing they were the chosen people, selected by God the Father to become a holy nation of priests that delivers the guide of a moral existence to the nations of the world. The peak-testosterone culture of ancient Israel was self-oriented. It was based in a specific territory and self-identity with a defined structure of laws and national borders, not seeking expansion to foreign lands, which were associated with pagan nations. However, with the decline in testosterone levels, the Hebrew people fell down in submission and were conquered and raped by multiple barbarian invasions, culminating in the final collapse of the Second Temple to the Roman Empire in 70 CE. This evolutionary process of one nation being overtaken by a greater empire gave rise to Christianity and eventually the Replacement Theology of Roman Catholicism, placing itself as the new chosen people, selected by God. In contrast to ancient Israel, the Roman Empire was concerned with the conquest of a multi-ethnic, multicultural, expanding empire, ruling over many peoples, and unifying humanity under an all-encompassing world order. This territorial expansion of the human social superorganism resulted in the synthesis of Judeo-Christian theology based on the Hebrew Bible and ancient Greek philosophy.

From the evolution of our Judeo-Christian heritage, we can ascertain that our view of our own social group, its boundaries in relation to other groups and to nature itself, is directed by sex hormones that frame our minds' perceptions of our place and position in the world. Religion is a cultural system of behaviors and practices, worldviews, ethics, and social organization that relate humanity to an order of existence. Testosterone increase caused the Protestant Reformation, which was a protest against the centralized control of the Catholic Pope. It was a trend toward more individualism and independence. Subsequently, testosterone collapse brought the Counter-Reformation back to Catholicism, disguised as the philosophy of secular humanism. Testosterone collapse in 1837 fomented feminism, altruism, Marxism, and Darwinism during the recessionary period of 1839-1859. In that year, the term *feminism* was coined by the French philosopher and utopian socialist, Charles Fourier. Another French socialist philosopher, Auguste Comte, coined the term *altruism* in the late 1830s, meaning "unselfishness," "the opposite of egoism," from "alter," or "other." Comte also founded the Religion of Humanity, later called secular humanism, a secular derivation of Catholicism, whose adherents must be emissaries of altruism. Comte came to believe in a utopian vision where feminine values are the embodiment and triumph of both sentiment and morality. Furthermore, during this time, the British Empire and the Chinese fought two Opium Wars over free trade and China's sovereignty. These wars led to drug legalization and mass addiction of the Chinese population. The First Opium War took place between 1839 and 1842, and the Second Opium War took place from 1856 to 1860.

## Man versus Mother Nature

The selfish-gene theory in biology, whose famous advocate is Richard Dawkins, states that genes are self-replicating machines that propagate themselves in self-interested survival. However, during evolution genes group together to form larger entities—the genetic sequences of DNA—that are the coding mechanisms in all living cells. In this process, individual genes relegate their independence in order to enjoy the benefits of becoming part of larger structures with greater complexity and hence the ability to further their survival. Evolution is the process through which nature creates by recombining genes to form new biological entities with different adaptations that are then selected through the struggle for the survival of the fittest. Random mutations cause changes in genes. Sex is the method that evolved to recombine genes in complex eukaryotic species. If females only replicated themselves in an asexual manner, no mixing of genes would be achieved, and the species would not evolve through recombination. To evolve–which means to change–the old must be destroyed to create the new; and it does this through the reproductive life cycle, which requires the gradual sacrifice of the parents to give birth to the new organism, their offspring.

Even the social groups we form are temporary, as one social dominance hierarchy eventually collapses and is taken over and replaced by another in order for the genes of the species to continue mixing and producing new forms. Both biological and cultural evolution is driven by this reproductive life cycle, programmed into our genetic makeup and regulated by our sex hormones. Each individual is fundamentally composed of a group that is a collection of smaller entities, such as nations made of individual people, people made of cells, cells made of organelles, and so forth, in ever decreasing hierarchical order. Evolutionary selection will always seek to mix the genes of individuals in each group together, even at the expense of sacrificing them, to achieve variation and to better adapt to changing environmental conditions. Dominance hierarchies, which give power to certain males, are simply a method evolution uses to create a transitory social structure that will be replaced according to changing levels of the dominance and status hormones.

As Stephen Jay Gould and Niles Eldredge stated in the theory of punctuated equilibrium during social turmoil of the early 1970s, periods of equilibrium and stability are fleeting. By human measurement, they might seem prolonged, yet inevitably they will be punctuated by sudden cataclysmic events that trigger great and sudden—often catastrophic—transformations in evolution. Hence, during the course of evolution, in order to adapt to changing environmental conditions, selfish genes had to experience phase transitions in their behavior through epigenetic switching, relinquishing their independence in order to coalesce into larger groups of genes to create more complex and adaptive biological structures. Those "unselfish" genes became the ancestors of complex, multi-cellular life, such as the human species. This evolutionary process of transition explains the success of our dual nature as a homo duplex, capable of shifting between selfish and group behavior by epigenetic adaptation to changing environments.

Man, through the testosterone regulating his life cycle, grows up to maturity to become fertile in order to reproduce. He might feel dominant, independent, and secure for a while, building families, social bonds of cultural significance, nations and civilizations, but his triumph is only for a season, and his euphoria fleeting and transient. Soon, his time too shall pass, and Mother Nature will demand that he surrenders himself back to her. As written in Genesis 3:19, "For dust thou art, and unto dust shalt thou return." Man struggles with nature to tame her, to impose order from chaos, and to dominate her through science, art, and industry. He forms social structures and organizations to develop civilization and seeks a woman to build a family that will provide meaning and purpose to life. In the end, however, he will submit to death. Nevertheless, his quest is not in vain, because his biological progeny and cultural heritage will live on through his contribution to the human species.

## Philosophy of Mind

To bring humanity into a greater understanding of its mammalian brain and to transcend biological inheritance, we must use our faculty of reason to command the animal within us. Mind–body dualism is the origin of our philosophical conundrum. It is a manifestation of Homo sapiens' attempt to bridge the gap between our evolutionary roots in instinctual animal life and our further evolution. We evolved to have a large brain and to be capable of high-level cognition, forming complex culture, social organization, and philosophical systems. In our struggle to assert our free will and independence of mind, in a revolt against the limitations imposed by our material body, we have built our Western philosophical worldview on the premise that we can disconnect the mind from the body in order for reason to gain control and master the will.

There are three main schools of thought in the Western tradition regarding this fundamental view on the philosophy of mind:

1. The blank slate hypothesis states that man is a perfectly rational being in full command of his senses and that he commands nature. This view is implicitly proclaimed in the most optimistic, peak-testosterone philosophy of the Hebrew Bible, which contends that man is above nature and the rest of animal world, made in the image of the Creator for the good of man. This approach is "this-world" oriented, and is also expounded on by prodigious thinkers, such as Aristotle, John Locke, and Ayn Rand. That which is rational, real, and pleasurable is the basis of the capitalist system created by the Founding Fathers of the United States, grounded in the biblical worldview during the period of the Enlightenment.

2. The utopian view states that man has the faculty of reason, but denies reality and seeks to escape existence. This less optimistic, Platonic worldview still regards man as a rational being, but seeks to escape the reality of this world, associated with sin and evil, to the realm of perfect ideas not affected by natural phenomena. This approach appeals to communist ideology that seeks to create a utopian, egalitarian society guided by the "rational" principles set by the philosopher king, but denies human free will in this world. It was the basis of Soviet communism and is based on feminine, egalitarian ideals compelled by testosterone decline. In the utopian model, man attempts to escape the bleak reality of existence and to seek happiness in another dimension of collective delusions.

3. Nihilism is the direly pessimistic view of human nature, stating that man has no free will, denying any capacity for reason or knowledge of objective reality or a rational basis for ethics. The famous, modern proponent of this philosophy, also called existentialism, was Nietzsche, who declared that God is dead; man is solely motivated by the will to power and is beyond good and evil. The German Nietzsche, who is heir to the Lutheran, Prussian, militaristic tradition of discipline and power, was also obsessed with animal

instincts for dominance and a hierarchical structure of society. This fascist ideology glorifies war waged by the dominant group, or master race, for power over the inferior groups marked as slaves. It was embraced by the National Socialists in Germany under Hitler. He and his regime rose to power in the sadomasochistic, homosexualized masculine culture that was based on butch–femme dominance hierarchies of Weimer Germany in a period of testosterone collapse. Not only denying the existence of happiness, this malevolent view demands man to be in a state of total war, a consummate Darwinian struggle for the survival of fittest. Such a nihilistic philosophy in a culture of depression was also the basis of the Dark Age Augustinian doctrine of Original Sin as well as the Islamic jihad of sexual submission, promising martyrs who die for the cause of holy war seventy-two virgins in heaven.

## Phase Transitions in Cultural Evolution

To continue with a review of the material addressed in this book, we can see how cultural phase transitions played out in the early history of Western society. The means of sex for mixing genes between different cultures and gene groups to create increased genetic variation evolved into cyclical phases.

*Inbreeding phase:*
The high-testosterone, patriarchal culture keeps the society in check and safeguards women from foreigners, protecting its members from mixing cultures and genes with others, in contrast to the communist free-sex culture that mixes genes freely between every group.

*Evolution:*
Evolution works by creating variations using sex as a means to reproduce new creatures that can better adapt to new and different environmental conditions, increasing their chances of survival and reproduction.

# 10 • TRANSCENDING DETERMINISM

*Outbreeding phase:*

When testosterone levels fall, the dominant Western male loses his dominance, and the feminists castrate him and bring in the barbarians to rape, violate, and inseminate them to produce cross-cultural genetic variation.

*The role of religion:*

We see this historical process in the peak-testosterone worldview evident in the period of the Hebrew Bible, in which the culture is initially extremely hostile to pagan—meaning foreign—influence and largely refrains from mixing with the culture and genes of other societies. However, as testosterone levels decline in the population, the patriarchy loses its dominance over the females and weakens due to internal loss of cohesion and group loyalty. This loss of the alpha male status enables barbarian forces to conquer and take over the society and to interbreed with its women, mixing both genes and ideas in a multicultural settings.

This process of cultural degeneration eventually led to the split of ancient Israel into two kingdoms and the conquest of the Kingdom of Israel by the Assyrians in 730 BCE. Later the Babylonians conquered the Kingdom of Judea in 597 BCE, causing the fall of the First Temple. With the fall of the Second Temple in 70 CE by the Romans, the Jews became assimilated into the Roman Empire, which gave rise to the new religion of Christianity. According to scientists, the origin of Ashkenazi Jews was when male Jews married Italian women as early as the Hellenistic period, in which Greek culture spread throughout the Middle East and influenced a Hellenistic Judaism following the conquests of Alexander the Great in the Fourth Century BCE. Hence, testosterone decline led to the loss of Jewish independence and the eventual spread of Christianity, an amalgam of Jewish ideas and Greek philosophy, across the nations of the Roman Empire.

## A New Philosophy of Mind for the Transcendent Man

Having analyzed the evolution of Western philosophy, three primary views of the human mind were presented: the blank slate, utopian view, and nihilism. These philosophies of mind are a product of different periods in history with altering social moods driven by hormone cycles that affect our cultural evolution. We now find ourselves in an age of significant scientific progress. Previously unimaginable technological advances, such as AI and genetic engineering, promise to allow humanity to achieve an evolutionary leap forward in rapidly accelerating time frames and thereby transcend our biology. We seem to stand either on a threshold of greatness or on the edge of a precipitous collapse.

Ray Kurzweil, Google's Director of Engineering and famed futurist, continues to maintain that a fusion between AI and humanity is only years away. Calling it the Singularity, he predicts that carbon- and silicon-based intelligence will meld, forming a single global consciousness: "By 2029, computers will have human-level intelligence," Kurzweil said. "That leads to computers having human intelligence, our putting them inside our brains, connecting them to the cloud, expanding who we are." His expectation, as well, is that machines could be smarter than the humans who created them.[15]

The burgeoning of the information age has taken place since the 1940s with the advances in AI, during a period of positive social mood propelled by increasing testosterone levels into the 1950s. More recently, Frank Rosenblatt created the perceptron in 1958, an electronic artificial neural network, trying to imitate the brain and exhibiting an ability to learn and classify objects. This model was later improved in the 1980s, leading to advancements in image and speech recognition. In recent years, Google's company DeepMind has used reinforcement learning, a machine-learning approach that mimics the human dopamine reward system to reach a goal. This has advanced AI, but there is still much to learn, because machines still lack any understanding of their physical environment, as psychologist Gary Marcus writes:

# 10 • TRANSCENDING DETERMINISM

Artificial Intelligence is colossally hyped these days, but the dirty little secret is that it still has a long, long way to go. Sure, A.I. systems have mastered an array of games, from chess and Go to "Jeopardy" and poker, but the technology continues to struggle in the real world. Robots fall over while opening doors, prototype driverless cars frequently need human intervention, and nobody has yet designed a machine that can read reliably at the level of a sixth grader, let alone a college student. Computers that can educate themselves — a mark of true intelligence — remain a dream.[16]

However, history demonstrates that our quest for intelligence suffers setbacks when social mood turns negative. These are called AI winters, periods when faith in its progress is weak and funding of research dries up. Two such periods were 1974-1980, in response to a report to the UK Parliament by mathematician James Lighthill that excoriated the state of AI progress, and 1987-1993, when the desktop computer prevailed over more expensive AI systems.[17] Both of these time frames coincided with testosterone collapse that caused a decline in reasoned motivation. The public fear of malevolent AI is another factor that ebbs and flows with social mood; and it can be a sign of another rising wave of irrationality that will stifle human development.

Not only does the public find enhanced intelligence potentially terrifying, however. Testosterone collapse has its effects in the high-tech community, leading to fear, anxiety, and demonization of reason, science, and AI advances. The CEO of Tesla and SpaceX, Elon Musk, fears that our ability to prevent robots from destroying humanity is less than ten percent! He has warned audiences of his employees as well as high-level government officials that humans "should be extremely afraid" and puts his hope in government regulation of the field of AI.[18] Is Musk's concern as dire as it sounds? Professor Alan Winfield argues that we are not, in fact, destined to be faced with Frankenstein's out-of-control monster in the form of the AI singularity. For this to happen, he states, a series of "big ifs" would need to transpire:

> *If* we succeed in building human equivalent AI and *if* that AI acquires a full understanding of how it works, and *if* it then succeeds in improving itself to produce super-intelligent AI, and *if* that super-AI, accidentally or maliciously, starts to consume resources, and *if* we fail to pull the plug, then, yes, we may well have a problem. The risk, while not impossible, is improbable.[19]

It is understandable that the revolution in machine intelligence raises the fundamental question of who will be in control of our world, humans or robots. Artificial intelligence is not biological; robots do not have any internal drive to live and replicate like animals do, the organisms with self-replicating genes. However, software code *can* be programmed to self-replicate and evolve, which is already available with global internet connectivity. This raises the risk that we humans, particularly in an environment of low testosterone, will become submissive and can be enslaved by the robots. In a variation on David Levy's robot love, the co-founder of Apple, Steve Wozniak, even imagines a future where humans could be kept like pets by a dominant robot race.[20] However, even if robots were more intelligent, we will still be able to control them as long as we have the impetus to dominate nature. Our merge with artificial intelligence could simply be a tool to enhance human intelligence and capabilities. The word hormone means impetus, and testosterone impels human growth and sexual reproductive cycles, shaping our minds and social structure. Therefore, hormonal cycles will continue to govern our world. However, in light of this eventuality, it is imperative that we learn to control our hormonal cycles in order not to fall into submission to our own creation of artificial intelligence.

A genuine concern that looms over the prospect of controlling our future is that because of low testosterone levels in society, even the high-tech community—composed predominantly of male engineers—is now politically controlled by the feminist, cultural Marxist mindset. In the summer of 2017, Google engineer James

# 10 • TRANSCENDING DETERMINISM

Damore was fired for submitting an internal memo, in response to a company request for suggestions, that simply questioned their feminist policy and bias in employment as counter to biological sex differentiation factors. In a similar vein, in the birthplace of Enlightenment science, the British Royal Society (of which Isaac Newton had been president) awarded a prize for a feminist propaganda book called *Testosterone Rex*. In it the author suggests that men are dinosaurs whose time has passed, and masculinity is only a cultural, not biological, construct. These examples show how testosterone collapse leads to the demise of modern science and submission of the human mind to otherworldly delusions associated with Platonic mysticism. Since 2009 AI has been in a great boom. Likewise, digital crypto-currencies such as bitcoin powered by the new decentralized technology of the block chain have been flourishing. However, the manic trend might mark that we are getting closer to the peak of a bubble, whose implosion could bring another wave of pessimism and another AI winter, along with a coming economic Great Depression.

## Hope for the Future

The future looks uncertain. Despite the deterministic aspects of biology that limit our free will, however, philosopher Daniel Dennett, in the popular science book *Freedom Evolves* (2003) suggests with a measure of optimism that there can be free will and moral accountability in a deterministic world moved by physical causality. He adds an evolutionary twist to the compatibilist approach that seeks to settle the contrast between free will and biological determinism. He advocates that although actions may be pre-determined in the strictly physical sense, we have evolved greater abilities to respond to events in a fashion that promotes our life.

Dennett's ideas are probably aimed in the right direction. We must seek rational and scientific solutions to our long lasting disputes over the philosophy of mind. First and foremost, we have to recognize what the nature of ideas is that shapes our minds and how they evolve.

The ideas we hold dear in the form of language are information composed of symbols, alphabet letters that depict the structure of the organization of matter. As noted previously, this parallels the chemical periodic table, in which letters represent each element, differentiated according to the quantity of protons in its atom. In biology, the letters in the genes that form DNA are symbols for the genetic instructions on how to assemble life

From the physical world to biology and philosophy, our universe is shaped by matter. We understand the organization of matter using information as a symbolic language that is in a constant process of evolution according to energy flows. Energy drives phase transitions in the organization of matter, whether on the levels of atoms in physics, cells in biology, or ideas in human cultures, through changing hormone levels. Energy and matter will always evolve, and we cannot and should not try to stop evolution; nevertheless, we have the capacity to make decisions to take charge of our evolution and direct it to promote our well-being and growth. By artificially controlling the intensity of light in our environment we can consciously regulate our hypothalamus, the master hormone regulator, and take charge of hormones. Hence, we can epigenetically turn on genes that produce happy chemicals that raise our social mood and motivate us to be rational and productive and establish a philosophy of life that promotes our flourishing.

Studying the accumulated knowledge of the ages in philosophy, religion, and science can help to guide us in understanding our own human nature and how to best proceed as a species in our best interest to fulfill our full potential as rational human beings. By this endeavor we will raise ourselves above what natural cycles have intended for us. Rabbi Maimonides agrees with Aristotle that to reason we must have the strong will to use logic to reach a higher level of understanding that will ultimately bring us to know God, the infinite perfection. Hence, we should learn the truth of each school of thought regarding our philosophy of mind, as each view of the human mind is generated in cultural settings as a product of the ideas and social mood of its time. The blank slate hypothesis of absolute rationality and confidence in our sense perception is of great merit, but should not cause us to discard the study of evolutionary psychology, and to even consider Nietzsche's nihilistic perspective of the human animal as a beast of prey fighting for dominance and survival.

Only by comprehensive understanding and integrating the different schools of thought with a broad scientific perspective on the nature of the evolutionary processes, can we stand tall and reach for the heavens in philosophy and science. With the understanding of how biological life is programed to react to light/dark cycles with growth/decay hormonal cycles that shape our evolution, we can now control our hormone levels and change the course of history. But, with this ability lies great responsibility. We have the great task of learning the intricate two-way relationship between our minds, our biology, and how energy flows from the physical world to affect our brains and social mood. Hence, on the one hand, this complicates our study of science, as we now know how critical it is to understand the scientists' state of mind that shape their worldviews—the social mood, philosophy, and culture of their age—that enables us to build a psychological profile of their character. On the other hand, by taking these parameters into consideration, we can attain a great leap forward in our comprehension of the laws that govern the universe, biology, and our brains.

We stand here in a profound time, with the opportunity to take this knowledge to the next level and allow our minds to proceed with confidence to achieve greatness for our species. However, we must deal with the biology of our brains and provide them with the proper physical environment that will promote their functioning at their absolute best to further our lives.

The irony is that the only solution to keep our minds from being trapped in the biological brain is not the blank slate hypothesis that imagines the mind to be free from matter, but actually to figure out and reverse engineer how our brain is designed to work, thereby learning to control its machinations in a rational and life-promoting fashion. Evolution works through trial and error, and we might make mistakes along the way; but this is the evolutionary path that produced us. Knowledge provides the power of the rational man over nature. Let us put our knowledge to good use!

## The Code of Creation

The atheist scientist Howard Bloom published *The God Problem: How a Godless Cosmos Creates* in 2016, seeking the most fundamental questions of creation: "How does the cosmos do something it has long been thought that only gods could achieve? How does an inanimate universe generate stunning new forms and unbelievable new powers without a Creator? How does the cosmos create?"[21] While religion fills a central role in defining our social structure and worldview, we cannot ignore the nature of our biological existence while seeking to discover it. However heroic Bloom's motivation, the quest to discover the formula of creation will remain impotent and doomed to fail without recognizing that sex hormones drive evolutionary cycles. These hormones are foundational to biological and cultural complexity, making complex life possible with the rise of sexual reproduction in the life-cycle pattern. The study of evolution and complexity is a mixture of the hormonally-driven masculine and feminine brains: the masculine promotes linear thinking about evolution as the progress of creation under the Creator God who commands nature, while the feminine focuses on the cyclic patterns in nature, including the fatalistic submission to our destiny as determined by natural cycles.

In addition, it is necessary to understand the philosophy of language that is the basis of creation, as stated in Genesis 1 in ancient Hebrew Scripture. Here it is declared that the same process of creation responsible for the emergence of order out of chaos in the universe was responsible for human civilization to develop through the ascendency of the rational and ethical man, created in the image of the Creator. Therefore, perhaps a more apt title for Bloom's book would be *The God Principle: How Nature Creates Order and Complexity.*

The burgeoning science of complex, adaptive systems offers a new frontier for humanity's glorious journey to discover the secrets of creation and to come ever closer to knowing the benevolent Creator of the universe. In the biblical philosophy of language, this vision of the end-times is described in Daniel 12:4 as such: "Knowledge shall be increased." This book has sought to align with that vision, suggesting that greater human understanding is the key to conquering nature.

# ACKNOWLEDGEMENTS

I want to thank my family and my father, Jacob, for their support for my work on the book.

# ABOUT THE AUTHOR

Roy Barzilai is an independent scholar, who studied both Ayn Rand's philosophy of Objectivism, and Rivka Schechter's philosophy of language, rooted in the Hebrew Bible. The synthesis of Rand's Aristotelian philosophy and the biblical creed of ethical monotheism provides profound insights into the ideas that shaped the Western mind. By exploring the intellectual history of Western civilization, Roy seeks to reach a greater understanding of the human mind.

As a financial analyst for over a decade, Roy became aware of the herd mentality in financial markets. He studied the Wave Principle of Human Social Behavior and the new science of Socionomics, focusing on how change in social mood affects society, its ideas, philosophy, culture, and economy, and is thus the engine of history.

Roy holds undergraduate degrees from Tel Aviv University in Law, accounting, and computer science.

Other Titles by Roy Barzilai

THE TESTOSTERONE HYPOTHESIS: *How Hormones Regulate the Life Cycles of Civilization* (2015)

THE OBJECTIVE BIBLE: Western *Civilization's Struggle for Philosophic Liberation from a Herd-mentality and Pagan Mysticism* (2014)

# NOTES

## Preface

1 Tyler Durden, "Calexit: Record Number of Californians Support Secession, New Poll Finds," ZeroHedge, January 1, 2017, http://www.zerohedge.com/news/2017-01-24/calexit-record-number-californians-support-secession-new-poll-finds.

2 Chiara Palazzo, "'This is Not about Religion - This is about Terror and Keeping our Country Safe' - Donald Trump Defends Executive Order and Attacks the Media," *The Telegraph*, January 30, 2017, http://www.telegraph.co.uk/news/2017/01/30/not-muslim-ban-donald-trump-defends-executive-order-attacks/.

3 Kate Pickles, "Imam Tells Muslim Migrants to 'Breed Children' with Europeans to 'Conquer Their Countries' and Vows: 'We Will Trample Them Underfoot, Allah Willing,'" *Daily Mail Online,* September 18, 2015, http://www.dailymail.co.uk/news/article-3240295/Imam-tells-Muslim-migrants-breed-children-Europeans-conquer-countries-vows-trample-underfoot-Allah-willing.html.

4 Alex VanNess, "Linda Sarsour, Women's March Organizer and Fake Feminist," *Breitbart*, February 3, 2017, http://www.breitbart.com/national-security/2017/02/03/vanness-linda-sarsour-womens-march/.

5 Kate Connolly, "Angela Merkel Defends Germany's Refugee Policy after Attacks," *The Guardian*, July 28, 2016, https://www.theguardian.com/world/2016/jul/28/merkel-rejects-calls-to-change-germanys-refugee-policy-after-attacks.

6 "German Government sets up Website to Tell 'Migrants' How to Have Sex with German Women," *Diversity Macht Frei* (blog), March 6, 2016, http://diversitymachtfrei.blogspot.co.il/2016/03/german-government-sets-up-website-to.html?m=1.

7 Tom Batchelor, "Merkel has not been Deterred in her Immigration Policy by this Wave of Merkel Admits EU is Being RAVAGED BY TERROR but says Germany Should Welcome MORE Refugees," *Express*, July 28, 2016, http://www.express.co.uk/news/world/694243/German-Chancellor-Angela-Merkel-terror-attacks-wont-stop-Germany-welcoming-refugees.

# Sex Wars

## Introduction

[1] Dana R. Carney, Amy J.C. Cuddy, and Andy J. Yap, "Power Posing: Brief Nonverbal Displays Affect Neuroendocrine Levels and Risk Tolerance," *Psychological Science* 21, no. 10 (2010): 1363–1368, doi: 10.1177/0956797610383437.

[2] Peter Wehrwein, "Astounding Increase in Antidepressant use by Americans," *Harvard Health* (blog), October 20, 2011, http://www.health.harvard.edu/blog/astounding-increase-in-antidepressant-use-by-americans-201110203624.

[3] Tom Curry, "The Evolution of Obama's Stance on Gay Marriage," *NBC News*, May 9, 2012, http://nbcpolitics.nbcnews.com/_news/2012/05/09/11623172-the-evolution-of-obamas-stance-on-gay-marriage?lite.

[4] Steve Marche, "The Unexamined Brutality of the Male Libido," *New York Times*, November 25, 2017, https://www.nytimes.com/2017/11/25/opinion/sunday/harassment-men-libido-masculinity.html?smprod=nytcore-ipad&smid=nytcore-ipad-share.

[5] Laura Kipnis' Facebook page, accessed November 26, 2017, https://www.facebook.com/laura.kipnis.

[6] Ibid.

[7] *Wikipedia*, s.v. "Islamic State of Iraq and the Levant," last modified January 6, 2018, https://en.wikipedia.org/wiki/Islamic_State_of_Iraq_and_the_Levant.

[8] Aljo, "Are Socialists Sadists?" *The Socialist Standard* no. 1325 (January 2015), http://www.worldsocialism.org/spgb/socialist-standard/2010s/2015/no-1325-january-2015/are-socialists-sadists.

[9] Vittorio Hernandez, "Japan's Population Time Bomb Worsens as 'Herbivore' Men not Interested in Sex Extends to Married Males," *International Business Times,* August 21, 2015, AU edition, http://www.ibtimes.com.au/japans-population-time-bomb-worsens-herbivore-men-not-interested-sex-extends-married-males-1461688.

[10] Joan Kaufman, "China now has the Lowest Fertility Rate in the World," *The Buzz* (blog), *The National Interest*, December 1, 2016, http://nationalinterest.org/blog/the-buzz/china-now-has-the-lowest-fertility-rate-the-world-18570.

[11] Eric Baculinao, "China Tackles 'Masculinity Crisis,' Tries to Stop 'Effeminate' Boys," January 9, 2017, *NBC News*, http://nbcnews.to/2jtLofS.

[12] Matt Blake, "German Population Plummets as Quarter of Men say 'No' to Kids," *Daily Mail*, August 21, 2013, http://www.dailymail.co.uk/news/article-2398796/German-population-shrinks-QUARTER-men-say-kids.html#ixzz4TTBu3kQq.

[13] Associated Press, "Swedish Government Wants to Find out if Swedes are Having Less Sex," *The Guardian*, July 29, 2016,

https://www.theguardian.com/world/2016/jul/29/sweden-government-study-sex-decline.

[14] Ingrid Carlqvist and Lars Hedegaard, "Sweden: Rape Capital of the West," *Gatestone Institute*, February 14, 2015, https://www.gatestoneinstitute.org/5195/sweden-rape.

[15] Valerie Hudson, "Europe's Man Problem," *Politico*, January 5, 2016, http://www.politico.com/magazine/story/2016/01/europe-refugees-migrant-crisis-men-213500.

[16] "Video: Egypt 1958, Muslims Laughed at Muslim Brotherhood Idea to Impose Hijab on Women: 'Let him (MB) wear it,'" *The Muslim Issue*, February 2, 2016, https://themuslimissue.wordpress.com/2016/02/02/video-egypt-1958-muslims-laughed-at-muslim-brotherhood-idea-to-impose-hijab-on-women/.

[17] Raja Halwani, "Why Sexual Desire is Objectifying – and Hence Morally Wrong," *Aeon*, December 9, 2016, https://aeon.co/ideas/why-sexual-desire-is-objectifying-and-hence-morally-wrong.

[18] Thomas Lemke, *Biopolitics: An Advanced Introduction* (NYU Press, 2011), 9-10.

[19] Laurette T. Liesen and Mary Barbara Walsh, "The Competing Meanings of 'Biopolitics,' in Political Science: Biological and Post-Modern Approaches to Politics," (paper presented at the American Political Science Association's Annual Meeting, Seattle, WA, September 1, 2011), 3.

# Chapter 1

[1] Camille Paglia, *Sexual Personae: Art and Decadence from Nefertiti to Emily Dickinson* (Yale Nota Bene, 2001), 3.

[2] Ibid, 2.

[3] "Health and the Environment: Testosterone Levels Fall Worldwide," *Washington's Blog, Global Research*, April 4, 2012, http://www.globalresearch.ca/health-and-the-environment-testosterone-levels-fall-worldwide/30129.

[4] James Dabbs, *Heroes, Rogues, and Lovers: Testosterone and Behavior*, (New York: McGraw-Hill, 2001).

[5] Descartes, René, *The Treatise on Man* in *The World and Other Writings*, ed. Stephen Gaukroger (Cambridge: Cambridge University Press 1662/1998), 202.

[6] António Damásio, *Descartes' Error: Emotion, Reason, and the Human Brain* (New York: Penguin, 2005).

[7] *Wikipedia*, s.v. "Somatic marker hypothesis," last modified March 8, 2015, https://en.wikipedia.org/wiki/Somatic_marker_hypothesis.

[8] Murray Gell-Mann, *The Quark and the Jaguar* (New York: St. Martin's Griffin, 1995).

[9] Michael Tomasello et al., "Understanding and Sharing Intentions: The Origins of Cultural Cognition," in *Behavioral and Brain Sciences* (Cambridge: Cambridge University Press 2004), 1, http://www.eva.mpg.de/psycho/staff/carpenter/pdf/Tomasello_et_al_inPressB BS.pdf.

[10] Ibid.

[11] Esther Herrmann et al., "Humans Have Evolved Specialized Skills of Social Cognition: The Cultural Intelligence Hypothesis," *Science* 317, no. 5843 (September 2007): 1360-1366, doi:10.1126/science.1146282.

[12] Michael Shermer, "Evolution Explains Why Politics Is So Tribal," *Scientific American*, June 1, 2012, https://www.scientificamerican.com/article/evolution-explains-why-politics-tribal/.

[13] Tessa Kendall, "Behaving Like Animals: *The Bonobo and the Atheist*," *The Guardian*, April 26, 2013, US edition, https://www.theguardian.com/science/the-lay-scientist/2013/apr/26/1.

[14] Jonathan Haidt, "Religion, Evolution, and the Ecstasy of Self-Transcendence," (lecture, Vancouver, BC, February 28, 2012) TED, https://www.ted.com/talks/jonathan_haidt_humanity_s_stairway_to_self_tran scendence/transcript.

[15] Jonathan Haidt, "Have We Evolved to Be Religious?" *Time*, Health & Science, March 27, 2012, http://ideas.time.com/2012/03/27/have-we-evolved-to-be-religious/.

[16] Haidt, "Religion, Evolution, and the Ecstasy of Self-Transcendence."

[17] "Religious Beliefs Activate Neural Reward Circuits in Same Way as Sex and Drugs," Neuroscience News, November 29, 2016 http://neurosciencenews.com/neurotheology-mpfc-reward-5622/.

[18] Mark Granovetter, "The Strength of Weak Ties," *American Journal of Sociology* 78 (1973): 1378.

[19] Walter A. Shelburne, Mythos and Logos in the Thought of Carl Jung: The Theory of the Collective Unconscious in Scientific Perspective (Albany, NY: SUNY Press, 1988), 4.

[20] Carl Jung, *Aspects of the Feminine* (Princeton: Princeton University Press, 1982), 65.

[21] Paglia, *Sexual Personae*, 40.

[22] *Wikimedia Commons*, s.v. "God the Father," Image, last updated June 1, 2017, https://commons.wikimedia.org/wiki/File:Cima_da_Conegliano,_God_the_Fa ther.jpg.

[23] *Wikimedia Commons*, s.v. "Schadow Grabmal Alexander," last modified November 19, 2017, https://commons.wikimedia.org/wiki/File:Schadow_Grabmal_Alexander_2.jpg

[24] Terry Teachout, "Siding with the Men," *New York Times*, July 22, 1990, Sunday Book Review, http://www.nytimes.com/1990/07/22/books/siding-with-the-men.html?pagewanted=all.

[25] Paglia, *Sexual Personae*, 37.

[26] Stephen L. Petranek, "American History Interviews Psychologist Jonathan Haidt," August 1, 2012, HistoryNet, http://www.historynet.com/american-history-interviews-psychologist-jonathan-haidt.htm.

[27] Haidt, Jonathan, *The Righteous Mind: Why Good People Are Divided by Politics and Religion* (New York: Pantheon, 2012). Kindle edition, 913.

[28] Paglia, *Sexual Personae*, 4.

[29] Steven Pinker, "Sex Ed," *New Republic*, February 14, 2005, https://newrepublic.com/article/68044/sex-ed.

[30] Steven Pruett-Jones, "The Ideological Opposition to Biological Truth," *Why Evolution is True* (blog), December 14, 2016, https://whyevolutionistrue.wordpress.com/2016/12/14/the-ideological-opposition-to-biological-truth/.

[31] Virginia Hale, "YouGov Poll: Only Two Percent of Men Aged 18-24 Feel Masculine," Breitbart London, May 20, 2016, http://www.breitbart.com/london/2016/05/20/yougov-masculine-poll/.

[32] Christine Hsu, "Is Testosterone the New Truth Serum? Male Sex Hormone Found to Promote Honesty in Men?"
October 10, 2012, Medical Daily, http://www.medicaldaily.com/testosterone-new-truth-serum-male-sex-hormone-found-promote-honesty-men-243038.

[33] Ibid.

[34] Stuart Taylor Jr., "Why Feminist Careerists Neutered Larry Summers," *The Atlantic*, February 2005, https://www.theatlantic.com/magazine/archive/2005/02/why-feminist-careerists-neutered-larry-summers/303795/.

[35] Sarah Lee, "Report: Science Says Women aren't as Capable as Men in Combat," The Blaze, December 11, 2016, https://www.theblaze.com/news/2016/12/11/report-science-says-women-arent-as-capable-as-men-in-combat/?utm_content=bufferbcfd2&utm_medium=social&utm_source=facebook.com&utm_campaign=buffer.

[36] Ibid.

[37] David Brooks, "The Limits of Empathy," *New York Times*, Opinion, September 29, 2011, http://www.nytimes.com/2011/09/30/opinion/brooks-the-limits-of-empathy.html.

[38] Paul Bloom, "The Perils of Empathy," *Wall Street Journal*, December 2, 2016, http://www.wsj.com/articles/the-perils-of-empathy-1480689513.

[39] Ibid.

[40] Paglia, *Sexual Personae*, 392.

[41] Carolyn Merchant, *The Death of Nature: Women, Ecology and the Scientific Revolution* (New York: Harper Collins, 1990, reprint), 278.

[42] Ibid., 282.

[43] Paglia, *Sexual Personae*, 12.

[44] Danielle Paquette, "Your Manliness Could be Hurting the Planet," *Washington Post*, Wonkblog, August 31, 2016, https://www.washingtonpost.com/news/wonk/wp/2016/08/31/your-manliness-could-be-hurting-the-planet/?utm_term=.fab406dba216.

[45] Ibid.

[46] Sam Dorman, "Paglia: 'Transgender Mania' is a Symptom of West's Cultural Collapse," CNS News, November 3, 2015, http://cnsnews.com/news/article/sam-dorman/camille-paglia-transgender-mania-symptom-cultural-collapse.

[47] Ibid.

[48] Ibid.

[49] Ibid.

[50] Jason Tucker and Jason VandenBeukel, "We're Teaching University Students Lies' – An Interview with Dr. Jordan Peterson," C2C Journal, December 1, 2016, http://www.c2cjournal.ca/2016/12/were-teaching-university-students-lies-an-interview-with-dr-jordan-peterson.

[51] Madison Park and Kiki Dhitavat, "Thailand's New Constitution Could Soon Recognize Third Gender, CNN, January 16, 2015, International Edition, http://edition.cnn.com/2015/01/16/world/third-gender-thailand/.

[52] Tucker and VandenBeukel, Peterson Interview.

[53] "High Rates of Suicide and Self-harm among Transgender Youth," Science Daily, August 31, 2016, https://www.sciencedaily.com/releases/2016/08/160831110833.htm.

[54] Tucker and VandenBeukel, Peterson Interview.

## Chapter 2

[1] Camille Paglia, *Free Women, Free Men: Sex, Gender, Feminism*. (New York: Pantheon, 2017), 6.

[2] Transliminal Media, "Religion, Myth, Science, Truth: An Evening of Darwinian Thought with Dr. Jordan B. Peterson." YouTube video, 2:35:31. Posted [November, 2015]. https://www.youtube.com/watch?v=o7Ys4tQPRis.

[3] Geoffrey Miller, *The Mating Mind: How Sexual Choice Shaped the Evolution of Human Nature* (New York: Anchor, 2001).

[4] "The Biology of Beauty," *Newsweek*, June 2, 1996, http://www.newsweek.com/biology-beauty-178836.

[5] Simon Baron-Cohen, "The Extreme Male Brain Theory of Autism," *Trends in Cognitive Sciences* 6, no. 6, (June 2002): http://cogsci.bme.hu/~ivady/bscs/read/bc.pdf.

[6] Alyson J. Lumley et al., "Sexual Selection Protects Against Extinction," *Nature* 522 (June 2015):470–473, doi:10.1038/nature14419.

[7] Lutz Becks and Aneil F. Agrawal, "The Evolution of Sex Is Favoured During Adaptation to New Environments, *PLOS*, May 1, 2012, https://doi.org/10.1371/journal.pbio.1001317.

[8] Root Gorelick and Henry H.Q. Heng, "Sex Reduces Genetic Variation: A Multidisciplinary Review." *Evolution* 65 (2011): 1088–1098. doi:10.1111/j.1558-5646.2010.01173.x.

[9] Laurence R. Gesquiere, et al., "Life at the Top: Rank and Stress in Wild Male Baboons," *Science* 333, no. 6040 (2001): 357–360, doi: 10.1126/science.1207120.

[10] Martin N. Muller, Richard W. Wrangham, "Dominance, Cortisol and Stress in Wild Chimpanzees (*Pan troglodytes schweinfurthii*)," *Behavioral Ecology and Sociobiology* 55, no. 4 (February 2004): 332–340, http://link.springer.com/article/10.1007%2Fs00265-003-0713-1.

[11] Scott Creel, "Dominance, Aggression, and Glucocorticoid Levels in Social Carnivores," *Journal of Mammalogy* 86, no. 2 (April 2005): 255–264, https://doi.org/10.1644/BHE-002.1.

[12] Robert M. Sapolsky, Cortisol Concentrations and the Social Significance of Rank Instability among Wild Baboons," *Psychoneuroendocrinology* 17, no. 6 (November 1992): 701–709, doi: http://dx.doi.org/10.1016/0306-4530(92)90029-7.

[13] John T. Hogg and Stephen H. Forbes, "Mating in Bighorn Sheep: Frequent Male Reproduction via a High-risk 'Unconventional' Tactic," *Behavioral Ecology and Sociobiology* 41, no. 1 (1997): 33-48. https://doi.org/10.1007/s002650050361.

[14] Peter-Frank Röseler, Ingeborg Röseler, Alain Strambi, and Roger Augier, "Influence of Insect Hormones on the Establishment of Dominance Hierarchies among Foundresses of the Paper Wasp, *Polistes gallicus*," *Behavioral Ecology and Sociobiology* 15, no. 2 (1984): 133–142, https://doi.org/10.1007/BF00299381.

[15] *Wikipedia*, s.v. "Dominance Hierarchy," last modified December 3, 2017, https://en.wikipedia.org/wiki/Dominance_hierarchy.

[16] Ibid.

[17] T. H. Clutton-Brock and E. Huchard, "Social Competition and Selection in Males and Females," *Philosophical Transactions of the Royal Society B: Biological Sciences* 368, no. 1631 (2013): 20130074, doi: 10.1098/rstb.2013.0074.

[18] Christopher Wanjek, "Meat, Cooked Foods Needed for Early Human Brain," *Live Science*, November 19, 2012, https://www.livescience.com/24875-meat-human-brain.html.

19 "Breaking the Food Chain Myth," *Free from Harm*, January 26, 2015, http://freefromharm.org/common-justifications-for-eating-animals/breaking-food-chain-myth/.

20 Gerd Leonhard, "Future of Technology and Impact on HR and Management," Image, Retrieved from https://www.flickr.com/photos/gleonhard/9563051196/in/photostream/. Modified per Creative Commons license, https://creativecommons.org/licenses/by-sa/2.0/.

21 *Wikimedia*, s.v. "Human Evolution Scheme," last modified October 16, 2015, https://commons.wikimedia.org/wiki/File:Human_evolution_scheme.svg.

22 Donald H. Edwards and Edward A. Kravitzt, "Serotonin, Social Status and Aggression," *Current Opinion in Neurobiology* 7 (1997): 812-819, http://caspar.bgsu.edu/~courses/Reading/Papers/1997EdwKra.pdf.

23 Rami Tzabar, "Do Animals Fight Wars?" *BBC*, August 11, 2015, http://www.bbc.com/earth/story/20150811-do-animals-fight-wars.

24 *Wikipedia*, s.v. "Gombe Chimpanzee War," last modified November 27, 2017, https://en.wikipedia.org/wiki/Gombe_Chimpanzee_War.

25 Michael Bang Petersen, "The Ancestral Logic of Politics: Upper-Body Strength Regulates Men's Assertion of Self-Interest over Economic Redistribution," *Psychological Science* 24, no. 7 (2013): 1098-1103, https://doi.org/10.1177/0956797612466415.

26 Nick Chowdrey, "Can MDMA Make You Racist?" April 3, 2014, http://www.vice.com/read/can-mdma-make-you-racist.

27 Ibid.

28 *Wikipedia*, s.v. "Male Warrior Hypothesis," last modified November 28, 2017, https://en.wikipedia.org/wiki/Male_warrior_hypothesis.

29 Ibid.

30 "Barack Obama Criticised for 'Treasonous' Bow to Japanese Emperor, *Telegraph*, November 16, 2009, http://www.telegraph.co.uk/news/worldnews/barackobama/6580190/Barack-Obama-criticised-for-treasonous-bow-to-Japanese-emperor.html.

31 Jane Goodall, "The ABC's of Chimpanzee Behavior," *Lessons for Hope, Jane Goodall Institute*, last modified January 14, 2018, http://www.lessonsforhope.org/abc/show_description.asp?abc_id=43.

32 Kate Taylor, "Stand Up Straight: How Minding Your Posture can Boost Your Self Confidence," *Up Coaching* (blog), March 1, 2016 http://www.upcoaching.co.uk/up-coaching-blog-posts/self-confidence-presence/.

33 Jesse Singal and Melissa Dahl, "Here Is Amy Cuddy's Response to Critiques of Her Power-Posing Research," *The Cut* (blog), *New York Magazine*, September

30, 2016, http://nymag.com/scienceofus/2016/09/read-amy-cuddys-response-to-power-posing-critiques.html.

[34] Bill Warner interview by Jamie Glazov, *Front Page Magazine*, posted on *Political Islam*, April 23, 2008, https://www.politicalislam.com/the-two-kinds-of-dhimmis/.

[35] Ibid.

[36] Susan Heitler, "How Do Sex and Power Abuses Lead to Terrorism and War?" *Resolution not Conflict* (blog), *Psychology Today*, September 11, 2011, https://www.psychologytoday.com/blog/resolution-not-conflict/201109/how-do-sex-and-power-abuses-lead-terrorism-and-war.

[37] Tawfik Hamid, *The Roots of Jihad* (Top Executive Media, 2006), 54-56.

[38] *Wikipedia*, s.v. "Female Genital Mutilation," last modified January 12, 2018, https://en.wikipedia.org/wiki/Female_genital_mutilation.

[39] *Wikipedia*, s.v. "Eunuch," last modified January 8, 2018, https://en.wikipedia.org/wiki/Eunuch.

[40] *Encyclopedia Judaica*, s.v. "Castration," *Jewish Virtual Library*, last modified January 14, 2018, http://www.jewishvirtuallibrary.org/jsource/judaica/ejud_0002_0004_0_0405 6.html.

[41] Laura Geggel, "The Science Behind Hitler's Possible Micropenis," *Live Science*, February 23, 2016, http://www.livescience.com/53818-hitler-micropenis.html.

[42] *Wikipedia*, s.v. "The Sexuality of Adolf Hitler," last modified January 8, 2018, https://en.wikipedia.org/wiki/Sexuality_of_Adolf_Hitler.

[43] S. M. Pearcey, K. J. Docherty, J. M. Dabbs, Jr., "Testosterone and Sex Role Identification in Lesbian Couples," *Physiology & Behavior* 60, no. 3 (1996): 1033-5, https://www.ncbi.nlm.nih.gov/pubmed/8873288.

[44] Michael Shermer, "When Ideas Have Sex," *Scientific American*, June 1, 2010, https://www.scientificamerican.com/article/when-ideas-have-sex/.

[45] David Derbyshire, "Are all Short Men Little Napoleons?" *Daily Mail*, June 15, 2015, Science, http://www.dailymail.co.uk/sciencetech/article-3125722/Are-short-men-little-Napoleons-s-said-smaller-men-tend-chippy-aggressive-s-scientific-evidence.html#ixzz4LdcLkLGa.

[46] NCD Risk Factor Collaboration, "A Century of Trends in Adult Human Height," *eLife* 5:e13410, ed. Eduardo Franco, July 26, 2016, doi: 10.7554/eLife.13410.

[47] Dan Kopf, "Where are the Tallest People in the World? Height Stagnation in the 20th Century," *Priceonomics*, August 22, 2016, https://priceonomics.com/where-are-the-tallest-people-in-the-world/.

[48] Ibid.

[49] Ibid.

50 *Wikipedia*, s.v. "Master–slave Morality: Revision History," last modified January 12, 2018,
https://en.wikipedia.org/wiki/Master%E2%80%93slave_morality.

51 Douglas Main, "Becoming King: Why So Few Male Lions Survive to Adulthood," *Live Science*, November 27, 2013,
http://www.livescience.com/41572-male-lion-survival.html.

52 Michael Balter, "Monogamy May Have Evolved to Prevent Infanticide," *Science*, July 29, 2013, http://www.sciencemag.org/news/2013/07/monogamy-may-have-evolved-prevent-infanticide.

53 *Wikipedia*, s.v. "Infanticide (zoology)," last modified December 21, 2017, https://en.wikipedia.org/wiki/Infanticide_(zoology).

54 Associated Press, "When Child Sex isn't Rape: French to Set Age of Consent," *ABC News*, November 14, 2017,
http://abcnews.go.com/Health/wireStory/child-sex-rape-french-set-age-consent-51146035.

55 A. D. Stenstrøm, "Stigende incidensrate af skizofreni hos børn og unge [Rising Incidence Rates of Schizophrenia among Children and Adolescents]," *Ugeskrift for Laeger* 2, no. 172 [in Danish] (2010): 2131-5,
https://www.ncbi.nlm.nih.gov/pubmed/20670588.

56 Kathleen Doheny, "Autism Cases on the Rise; Reason for Increase a Mystery," *WebMD*, last modified January 14, 2018,
http://www.webmd.com/brain/autism/searching-for-answers/autism-rise.

57 Christopher Badcock, *The Imprinted Brain: How Genes Set the Balance Between Autism and Psychosis* (London: Jessica Kingsley Publishers, 2009).

58 Satoshi Kanazawa, "Male Brain vs. Female Brain II: What is an 'Extreme Male Brain'? What is an 'Extreme Female Brain'?" *The Scientific Fundamentalist* [blog], *Psychology Today*, March 21, 2008,
https://www.psychologytoday.com/blog/the-scientific-fundamentalist/200803/male-brain-vs-female-brain-ii-what-is-extreme-male-brain.

59 Jeremy Waldron, *God, Locke, and Equality: Christian Foundations in Locke's Political Thought* (Cambridge, UK: Cambridge University Press, 2002).

60 David S. FitzGerald and David Cook-Martín, "The Geopolitical Origins of the U.S. Immigration Act of 1965," *Migration Policy Institute*, February 5, 2015,
http://www.migrationpolicy.org/article/geopolitical-origins-us-immigration-act-1965.

61 R. B. Slatcher, P. H. Mehta, and R. A. Josephs, "Testosterone and Self-Reported Dominance Interact to Influence Human Mating Behavior," *Social Psychological and Personality Science* (2011), doi: 10.1177/1948550611400099.

[62] P. H. Mehta, and R. A. Josephs, "Testosterone and Cortisol Jointly Regulate Dominance: Evidence for a Dual-hormone Hypothesis," *Hormones and Behavior* 58, no. 5 (2010):898-906, doi: 10.1016/j.yhbeh.2010.08.020.

[63] John Stonestreet, "The Death of Masculinity: Where Have All the Men Gone?" *CNS News*, September 13, 2016, http://cnsnews.com/commentary/john-stonestreet/death-masculinity-where-have-all-men-gone?utm_source=facebook&utm_medium=CNS&utm_term=facebook&utm_content=facebook&utm_campaign=c-masculinity-stonestreet.

[64] Roy Baumeister, "Cuckolding Fetish Relationships: Men Wanting Partners to Sleep with Other Men Reaches New High," *Independent*, November 18, 2016, http://www.independent.co.uk/life-style/love-sex/cuckold-fetish-sex-relationships-men-want-partners-cheat-new-high-a7423616.html.

[65] Heda Albeity, "Capturing Chinggis Qahan in the Secret History of Mongols," *The Blog of Heda Albeity*, April 22, 2010, http://albeityacademic.blogspot.co.uk/2010/05/capturing-chinggis-qahan-in-secret.html.

## Chapter 3

[1] Blaise Pascal, "Preface to the Treatise on Vacuum," *Minor Works*, trans. O. W. Wright, Vol. 48, Part 3, of 51, The Harvard Classics, ed. Charles W. Eliot (New York: P.F. Collier & Son, 1909–14), http://www.bartleby.com/48/3/10.html.

[2] J. C. D. Clark, ed., *Reflections on the Revolution in France*. A Critical Edition (Stanford: Stanford University Press, 2001), 261.

[3] Thomas Sowell, *A Conflict of Visions: Ideological Origins of Political Struggles* (Kent, UK: Quill, 1987), 162.

[4] Jesse Ferreras, "Millennials Like Communism More Than Their Parents Do. Capitalism? Not So Much," *The Huffington Post Canada*, October 18, 2016, Business, http://www.huffingtonpost.ca/2016/10/18/millennials-communism_n_12550502.html.

[5] Ibid.

[6] Anugrah Kumar, "Millennials Agree with Marx More than the Bible, Study Finds," *Christian Post*, October 29, 2016, http://m.christianpost.com/news/millennials-agree-with-marx-more-than-the-bible-study-finds-171191/?m=1.

[7] "Men's Testosterone Levels Predict Competitiveness," Medical Express, December 4, 2006, https://medicalxpress.com/news/2006-12-men-testosterone-competitiveness.html.

[8] Jocelyn Kiley and Michael Dimock, "The GOP's Millennial Problem Runs Deep," Pew Research Center, September 25, 2014,

http://www.pewresearch.org/fact-tank/2014/09/25/the-gops-millennial-problem-runs-deep/.

[9] Jasper Hamill, "Trigger Warning: Meet 'Generation Snowflake' – The Hysterical Young Women who can't Cope with Being Offended," *The Sun*, June 10, 2016, https://www.thesun.co.uk/news/1254262/meet-generation-snowflake-the-hysterical-modern-kids-who-cant-cope-with-being-offended/.

[10] Ibid.

[11] Amanda Prestigiacomo, "BuzzFeed Guys Test Their Testosterone Levels. The Results are Exactly What You'd Expect," *Daily Wire*, October 30, 2017, http://www.dailywire.com/news/22906/buzzfeed-guys-test-their-testosterone-levels-amanda-prestigiacomo#.

[12] Frederick Engels, Chapter 2, "The Family": Part 3, "The Pairing Family," *Origins of the Family, Private Property, and the State*, last modified October 14, 2010, https://www.marxists.org/archive/marx/works/1884/origin-family/ch02c.htm.

[13] Ibid.

[14] *Wikipedia*, s.v. "HebbianYerkesDodson," last modified November 29, 2014, https://commons.wikimedia.org/wiki/File:HebbianYerkesDodson.svg.

[15] David Gems, Linda Partridge, "Stress-Response Hormesis and Aging: 'That which Does Not Kill Us Makes Us Stronger'," *Cell Metabolism* 7, no. 3 (2008): 200-203, doi: http://dx.doi.org/10.1016/j.cmet.2008.01.001.

[16] S. J. Lupiena et al., "The Effects of Stress and Stress Hormones on Human Cognition: Implications for the Field of Brain and Cognition," *Brain and Cognition* 65, no. 3, (2007): 209-237, https://doi.org/10.1016/j.bandc.2007.02.007.

[17] Michael Kling, "Why Rich Families Lose Their Wealth," *NewsMax*, June 25, 2014, Finance, http://www.newsmax.com/Finance/StreetTalk/family-wealth-heir-children/2014/06/25/id/579236/.

[18] Ashley Montagu, *Man and Aggression* (New York: Oxford University Press, 1968) cited by Steven Pinker, *The Blank Slate: The Modern Denial of Human Nature* (New York: Penguin, 2002), 24.

[19] M. McGue, T. J. Bouchard, Jr., W. G. Iacono, and D. T. Lykken, "Behavioral Genetics of Cognitive Stability: A Life-span Perspectiveness. In R. Plominix and G. E. McClearn eds., *Nature, Nurture, and Psychology* (Washington, DC: American Psychological Association, 1993): 59-76.

[20] W. Durham, *Coevolution: Genes, Culture and Human Diversity*. (Stanford: Stanford University Press, 1991), Chapter 5.

[21] Dan Hurley, "Grandma's Experiences Leave a Mark on your Genes," *Discover*, May 13, 2013, Health and Medicine, http://discovermagazine.com/2013/may/13-grandmas-experiences-leave-epigenetic-mark-on-your-genes.

[22] Neil Howe, "Where did Steve Bannon get his Worldview? From my Book," *Washington Post*, February 24, 2017, Books,

https://www.washingtonpost.com/entertainment/books/where-did-steve-bannon-get-his-worldview-from-my-book/2017/02/24/16937f38-f84a-11e6-9845-576c69081518_story.html?utm_term=.cbd22c217f08.

23 Ibid.

24 Ibid.

25 Ibid.

26 Ibid.

27 Ibid.

28 *Holocaust Encyclopedia*, s.v. "Indoctrinating Youth" website of the United States Memorial Holocaust Museum, last modified January 16, 2018, https://www.ushmm.org/wlc/en/article.php?ModuleId=10007820.

## Chapter 4

1 Margaret Thatcher, Interview in *Women's Own*, 1987, https://www.theguardian.com/politics/2013/apr/08/margaret-thatcher-quotes.

2 "Karl Marx was Right, Socialism Works," Interview with E. O. Wilson at Harvard University, March 27, 1997, http://www.froes.dds.nl/WILSON.htm.

3 Martin A. Nowak, Corina E. Tarnita, and Edward O. Wilson, "The Evolution of Eusociality," *Nature* 466, no. 7310 (2010): 1057–1062, doi: 10.1038/nature09205.

4 Ibid.

5 Clint A. Penick, Colin S. Brent, Kelly Dolezal, and Jürgen Liebig, "Neurohormonal Changes Associated with Combat and the Formation of a Reproductive Hierarchy in the Ant *Harpegnathos saltator*," *Journal of Experimental Biology* (January, 2014), doi: 10.1242/jeb.098301.

6 Ibid.

7 Solenn Patalano, Timothy Hore, Wolf Reik, and Seirian Sumner, "Shifting Behaviour: Epigenetic Reprogramming in Eusocial Insects." *Current Opinion in Cell Biology* 24 (2012): 367-73, doi: 10.1016/j.ceb.2012.02.005.

8 Ibid.

9 "Karl Marx was Right, Socialism Works," Interview with E. O. Wilson.

10 Stephanie Pappas, "Why Pregnancy Really Lasts 9 Months," *Live Science*, August 27, 2012, http://www.livescience.com/22715-pregnancy-length-baby-size.html.

11 Christopher Opie, Quentin D. Atkinson, Robin I. M. Dunbar, and Susanne Shultz, "Male Infanticide Leads to Social Monogamy in Primates, *Proceedings of the National Academy of Sciences of the United States of America* 110 no. 33, Anthropology (August 2013): 13328–13332, doi: 10.1073/pnas.1307903110.

12 Ibid.

13 *Wikipedia*, s.v. "r/K Selection Theory," last modified December 28, 2017, https://en.wikipedia.org/wiki/R/K_selection_theory.

[14] Anonymous Conservative, *The Evolutionary Psychology Behind Politics: How Conservatism and Liberalism Evolved Within Humans* (Federalist Publications: 2017).

[15] Ibid.

[16] Ibid.

[17] Herbert Gintis, "Gene–culture Coevolution and the Nature of Human Sociality," *Philosophical Transactions of the Royal Society B: Biological Sciences* 366, no. 1566 (2011), doi: 10.1098/rstb.2010.0310.

[18] Ibid.

[19] Ibid.

[20] "A Brain Chemical Changes Locusts from Harmless Grasshoppers to Swarming Pests," website of the University of Cambridge, Research News, 30 Jan 2009, http://www.cam.ac.uk/research/news/a-brain-chemical-changes-locusts-from-harmless-grasshoppers-to-swarming-pests.

[21] Pete Wilton, "Locusts Pick Swarms as Lesser Evil," *Oxford Science Blog*, January 30, 2009, http://www.ox.ac.uk/news/science-blog/locusts-pick-swarms-lesser-evil.

[22] I. A. Barannikova, L. V. Baiunova, T. B. Semenkova, and A. B. Gruslova, ["Steroid Hormones in the Regulation of Migration in Fishes (Study of the Russian Sturgeon)"], *Rossiiskii Fiziologicheskii Zhurnal Imeni I.M. Sechenova / Rossiiskaia Akademiia Nauk* [Russian Journal of Physiology] 89, no. 11 [in Russian] (2003):1380-1387, https://www.ncbi.nlm.nih.gov/pubmed/14758663.

[23] Marilyn Ramenofsky and John C. Wingfield, "Regulation of Migration," *BioScience* 57, no. 2 (2007): 135-143, https://doi.org/10.1641/B570208.

[24] S. Nishizawa, "Differences Between Males and Females in Rates of Serotonin Synthesis in Human Brain," *Proceedings of the National Academy of Sciences of the United States of America* 94, no. 10, Medical Sciences (1997): 5308–5313, https://www.ncbi.nlm.nih.gov/pmc/articles/PMC24674/.

[25] Shasha Zhanga, Yan Liua, and Yi Raoa, "Serotonin Signaling in the Brain of Adult Female Mice is Required for Sexual Preference," *Proceedings of the National Academy of Sciences of the United States of America* 110, no. 24 (2013):9968–9973, doi: 10.1073/pnas.1220712110.

[26] Jonathan Haidt, "Moral Psychology and The Misunderstanding of Religion," *Edge*, September 21 2007, https://www.edge.org/conversation/moral-psychology-and-the-misunderstanding-of-religion.

[27] "Political Polarization in the American Public," *Pew Research Center*, U.S. Politics & Policy, June 12, 2014, http://www.people-press.org/2014/06/12/political-polarization-in-the-american-public/.

[28] *Wikipedia*, s.v. "Human Population Planning," last modified December 6, 2017, https://en.wikipedia.org/wiki/Human_population_planning.

[29] Carla Garnett, "Plumbing the 'Behavioral Sink,'" *NIH Record* LX, no. 15 (2008), https://nihrecord.nih.gov/newsletters/2008/07_25_2008/story1.htm.

[30] *Wikipedia*, s.v. "Blood and Soil," last modified December 20, 2017, https://en.wikipedia.org/wiki/Blood_and_Soil#Nazi_ideology.

[31] Michael E. Zimmerman, "Eco-Fascism," *Encyclopedia of Religion and Nature,* ed. Bron Taylor (New York: Continuum International Publishing Group, 2008): 531-532.

[32] *Wikipedia*, s.v. "Blood and Soil."

## Chapter 5

[1] Lynn Margulis, "Did Sex Emerge from Cannibalism? Sex, Death and Kefir. By Lynn Margulis (1938–2011)," Scientific American, November 23, 2011 [August 1994], https://www.scientificamerican.com/article/sex-death-kefir-lynn-margulis/.

[2] Ibid.

[3] Ibid.

[4] L. T. Morran, B. J. Cappy, J. L. Anderson, P. C. Phillips, "Sexual Partners for the Stressed: Facultative Outcrossing in the Self-fertilizing Nematode *Caenorhabditis elegans*," *Evolution* 63, no. 6 (2009):1473-82, doi: 10.1111/j.1558-5646.2009.00652.x.

[5] Richard E. Michod and Aurora Nedelcu, "Need Sex? It's Probably Something about Stress," *Eurek Alert!*, June 8, 2004, http://www2.unb.ca/vip/amnedelcu/pdfs/eurekalert.PDF.

[6] Ibid.

[7] William R. Clark, *Sex and the Origins of Death* (Oxford: Oxford University Press, 1998).

[8] Ibid.

[9] Nick Lane, Power, *Sex, Suicide: Mitochondria and the Meaning of Life* (Oxford: Oxford University Press, 2006).

[10] Joseph Schumpeter, *Capitalism, Socialism, and Democracy* (London: Routledge, 1994 [1942]): 182-83.

[11] *Wikipedia*, s.v. "Lynn Margulis," last modified January 9, 2018, https://en.wikipedia.org/wiki/Lynn_Margulis.

[12] C. Mann, "Lynn Margulis: Science's Unruly Earth Mother," *Science* 252, no. 5004 (1991): 378–381, doi: 10.1126/science.252.5004.378.

[13] Lynn Margulis, [interviewed in] *The Third Culture* by John Brockman (New York: Simon and Schuster, 1995), 133.

[14] John Brockman, "Lynn Margulis 1938-2011 'Gaia is a Tough Bitch,'" *Edge*, last updated January 21, 2017, https://www.edge.org/conversation/lynn_margulis-lynn-margulis-1938-2011-gaia-is-a-tough-bitch.

[15] Ibid.

[16] James A. Shapiro, *Evolution: A View from the 21st Century* (Upper Saddle River, NJ: FT Press, 2011)

[17] P. C. Davies, C. H. Lineweaver, "Cancer Tumors as Metazoa 1.0: Tapping Genes of Ancient Ancestors," *Physical Biology* 8, no. 1 (2011):015001, doi: 10.1088/1478-3975/8/1/015001.

[18] Yan Wang and Lloyd H. Kasper, "The Role of Microbiome in Central Nervous System Disorders," *Brain, Behavior, and Immunity* 38 (2014):1-12, doi: 10.1016/j.bbi.2013.12.015.

[19] Ibid.

[20] Scott F. Gilbert, Jan Sapp and Alfred I. Tauber, "A Symbiotic View of Life: We Have Never Been Individuals," *The Quarterly Review of Biology* 87, no. 4 (December 2012), pp. 325-341, http://www.jstor.org/stable/10.1086/668166.

[21] Ibid., 325.

[22] Ibid., 326.

[23] David Dobbs, "Die, Selfish Gene, Die," *Aeon*, December 13, 2013, https://aeon.co/essays/the-selfish-gene-is-a-great-meme-too-bad-it-s-so-wrong.

[24] Ibid.

[25] Ibid.

[26] Ibid.

[27] Ibid.

[28] Ibid.

[29] Ibid.

[30] Martin Brasier, "Is Efficiency Fatal?" *The Blog, Huffington Post*, August 29, 2012, http://www.huffingtonpost.com/martin-brasier/is-efficiency-fatal_b_1639369.html.

[31] Ibid.

[32] *Wikipedia*, s.v. "The Major Transitions in Evolution," last modified November 4, 2017, https://en.wikipedia.org/wiki/The_Major_Transitions_in_Evolution

[33] Richard E. Michod and Matthew D. Herron, "Cooperation and Conflict during Evolutionary Transitions in Individuality," *Journal of Evolutionary Biology* 19, no. 5 (2006): 1406, doi: 10.1111/j.1420-9101.2006.01142.x.

[34] Ibid.

[35] Ibid.

[36] "Watch Ball Bearings Organise Themselves into Complex Tree-like Structures," *Technology*, April 20, 2017, https://cosmosmagazine.com/technology/watch-ball-bearings-organise-themselves-into-complex-tree-like-structures.

[37] Douglas Medin, Carol D. Lee, and Megan Bang, "Point of View Affects how Science is Done," *Scientific American*, October 1, 2014, https://www.scientificamerican.com/article/point-of-view-affects-how-science-is-done/.

# Chapter 6

[1] Ayn Rand, "The Metaphysical and the Man-Made," *Philosophy: Who Needs It?* (Indianapolis, IN: Bobbs-Merrill, 1982), 26.

[2] Niall Firth, "Why Women Live Longer than Men: Male Bodies are Much More Genetically 'Disposable,'" *Daily Mail*, October 26, 2010, http://www.dailymail.co.uk/sciencetech/article-1323571/Why-women-live-longer-men-Male-bodies-genetically-disposable.html.

[3] Ibid.

[4] Yuriy Gorodnichenko and Gerard Roland, "Culture, Institutions and the Wealth of Nations," NBER Working Paper No. 16368, September 2010, http://eml.berkeley.edu/~ygorodni/gorrol_culture.pdf.

[5] Ibid., 2.

[6] Joan Y. Chiao and Katherine D. Blizinsky, "Culture–gene Coevolution of Individualism–Collectivism and the Serotonin Transporter Gene," *Philosophical Transactions of the Royal Society B: Biological Sciences* (2009), doi: 10.1098/rspb.2009.1650.

[7] Ibid.

[8] Baldwin M. Way and Matthew D. Lieberman, "Is There a Genetic Contribution to Cultural Differences? Collectivism, Individualism and Genetic Markers of Social Sensitivity," *Social Cognitive and Affective Neuroscience* 5, nos. 2-3 (2010): 203–211, https://doi.org/10.1093/scan/nsq059.

[9] Joan Y. Chiao and Katherine D. Blizinsky, "Culture–gene Coevolution."

[10] Ibid.

[11] Ibid.

[12] Ibid.

[13] M. J. Gelfand and A. Realo, "Individualism-collectivism and Accountability in Intergroup Negotiations," *Journal of Applied Psychology* 84, no. 5 (1999): 721-736, http://dx.doi.org/10.1037/0021-9010.84.5.721.

[14] Shinobu Kitayama, et al., "Dopamine-system Genes and Cultural Acquisition: The Norm Sensitivity Hypothesis," *Current Opinion in Psychology* 8 (2016):167–174, http://dx.doi.org/10.1016/j.copsyc.2015.11.006.

[15] Ibid.

[16] M. J. Gelfand and A. Realo, "Individualism-collectivism and Accountability."

[17] Ibid.

[18] Shinobu Kitayama, et al., "The Dopamine D4 Receptor Gene (DRD4) Moderates Cultural Difference in Independent Versus Interdependent Social Orientation," *Psychological Science* 25, no.6 (2014): 1169–1177, http://biosocialmethods.isr.umich.edu/wp-content/uploads/2015/03/Psychological-Science-2014-Kitayama.pdf.

19 Richard P. Ebstein, et al., "Association between the Dopamine D4 Receptor Gene exon III Variable Number of Tandem Repeats and Political Attitudes in Female Han Chinese," *Philosophical Transactions of the Royal Society B: Biological Sciences* 282, no. 1813 (2015), doi: 10.1098/rspb.2015.1360.

20 Georg S. Kranz, et al., "High-Dose Testosterone Treatment Increases Serotonin Transporter Binding in Transgender People," *Biological Psychiatry* 78, no. 8 (2014): 525–533, doi: 10.1016/j.biopsych.2014.09.010.

21 G. Schneider, et al., "Sex Hormone Levels, Genetic Androgen Receptor Polymorphism, and Anxiety in ≥50-year-old Males," *The Journal of Sexual Medicine* 8, no. 12 (2011): 3452-64, doi: 10.1111/j.1743-6109.2011.02443.x.

22 Meghan B. Oakes, Aimee D. Eyvazzadeh, Elisabeth Quint, and Yolanda R. Smith, "Complete Androgen Insensitivity Syndrome–A Review," *Journal of Pediatric & Adolescent Gynecology* 21, no. 6 (2008): 305-10, doi: 10.1016/j.jpag.2007.09.006.

23 M. Zitzmann, J. Gromoll, E. Neischlag, "The Androgen Receptor CAG Repeat Polymorphism," *Andrologia* 37, no. 6 (2005): 216, doi: 10.1111/j.1439-0272.2005.00692.x.

24 D. Milatiner, "Associations between Androgen Receptor CAG Repeat Length and Sperm Morphology," *Human Reproduction* 19, no. 6 (2004): 1426-30, doi: 10.1093/humrep/deh251.

25 H. R. Hull, et al., "Fat-free Mass Index: Changes and Race/Ethnic Differences in Adulthood," *International Journal of Obesity* 35, no. 1 (2011): 121-7, doi: 10.1038/ijo.2010.111.

26 J. M. Diamond, "Ethnic Differences: Variation in Human Testis Size," *Nature* 320 (1986): 488-489, doi: 10.1038/320488a0.

27 Christopher Ryan, Cacilda Jethá, *Sex at Dawn: The Prehistoric Origins of Modern Sexuality* (New York: Harper Collins, 2012), 241.

28 M. Wilson, M. Daly, "The Man who Mistook his Wife for a Chattel," in *The Adapted Mind*, ed. J. H. Barkow, L. Cosmides, and J. Tooby (New York: Oxford University Press, 1992), 299.

29 Koji Steven, "Do Asians Have the Smallest Testicle?" *8 Asians*, January 25, 2012, http://www.8asians.com/2012/01/25/do-asians-have-the-smallest-testicle/comment-page-1/.

30 Christopher Ryan, Cacilda Jethá, *Sex at Dawn*, 242.

31 Steven, "Do Asians Have the Smallest Testicle?"

32 Ibid.

33 See discussion at Quote Investigator: "You Can Avoid Reality, But You Cannot Avoid the Consequences of Avoiding Reality," *Quote Investigator*, April 30, 2015, https://quoteinvestigator.com/2015/04/30/reality/#return-note-11087-1.

[34] J. Philippe Rushton and Arthur R. Jensen, "Thirty Years of Research on Race Differences in Cognitive Ability," *Psychology, Public Policy, and Law* 11, no. 2, (2005): 265, doi: 10.1037/1076-8971.11.2.235.

[35] Ibid.

[36] Ibid., 266.

[37] Ibid.

[38] John Archer, Barbara Lloyd, *Sex and Gender* (Cambridge, UK: Cambridge University Press, 2002).

[39] Richard Lynn, "Sorry, Men ARE More Brainy than Women (and More Stupid too!) It's a Simple Scientific Fact, Says one of Britain's Top Dons," *Daily Mail Online*, May 8, 2010, http://www.dailymail.co.uk/debate/article-1274952/Men-ARE-brainy-women-says-scientist-Professor-Richard-Lynn.html.

[40] Gregory Cochran, Jason Hardy, and Henry Harpending, "Natural History of Ashkenazi Intelligence," *Journal of Biosocial Science* 38, no. 5 (2006): 659-93, doi: 10.1017/S0021932005027069.

[41] Kevin B. MacDonald, *A People That Shall Dwell Alone: Judaism as a Group Evolutionary Strategy, with Diaspora Peoples* (Bloomington, IN: iUniverse, 2002).

[42] Kevin B. MacDonald, "A Social Identity Theory of Anti-Semitism," in *Separation and Its Discontents: Toward an Evolutionary Theory of Anti-Semitism* (Santa Barbara, CA: Praeger, 1998), 1, http://www.kevinmacdonald.net/SAIDchap1.pdf.

[43] Ibid., 2.

[44] Nico Voigtländer and Hans-Joachim Voth, "Married to Intolerance: Attitudes towards Intermarriage in Germany, 1900-2006" (presentation, Annual Meeting of the Allied Social Science Associations, San Diego, CA, January 4-6, 2013), 10, https://www.aeaweb.org/conference/2013/retrieve.php?pdfid=316.

[45] Laurie Goodstein, "Poll Shows Major Shift in Identity of U.S. Jews," *New York Times*, October 1, 2013, http://www.nytimes.com/2013/10/01/us/poll-shows-major-shift-in-identity-of-us-jews.html?ref=us.

[46] M. Zitzmann and E. Nieschlag, "Testosterone Levels in Healthy Men and the Relation to Behavioural and Physical Characteristics: Facts and Constructs," *European Journal of Endocrinology* 144 (2001): 185, http://www.healthmegamall.com/Articles/BabeskinArticle224.pdf.

[47] Ibid.

[48] Dov Cohen, Richard E. Nisbett, Brian F. Bowdle, and Norbert Schwarz, "Insult, Aggression, and the Southern Culture of Honor: An 'Experimental Ethnography,'" *Journal of Personality and Social Psychology* 70, no. 5 (1996): 945-960, https://deepblue.lib.umich.edu/bitstream/handle/2027.42/92155/InsultAggressionAndTheSouthernCulture.pdf

49 Ibid., 949.

50 Ibid.

51 Allan Dafoe and Devin Caughey, "Honor and War: Southern US Presidents and the Effects of Concern for Reputation," *World Politics* 68, no. 2 (2016): 341-381, https://doi.org/10.1017/S0043887115000416.

52 Zitzmann and Nieschlag, "Testosterone Levels in Healthy Men," 185.

53 P. J. Henry, "Low-status Compensation: A Theory for Understanding the Role of Status in Cultures of Honor," *Journal of Personality and Social Psychology* 97, no. 3 (2009):451-66. doi: 10.1037/a0015476.

54 Howard Bloom, "Interview with Howard Bloom--Part I," interviewed by Grégoire Canlorbe, *Gatestone Institute*, November 28, 2016, https://www.gatestoneinstitute.org/9408/howard-bloom.

55 Howard Bloom, "Interview with Howard Bloom--Part II," interviewed by Grégoire Canlorbe, *Gatestone Institute*, November 28, 2016, https://www.gatestoneinstitute.org/10351/howard-bloom-interview-part-ii.

56 Ibid.

57 Ibid.

58 Ibid.

59 Ibid.

60 Ibid.

61 John Gray, "Steven Pinker is Wrong about Violence and War," *The Guardian*, March 13, 2015, Society, https://www.theguardian.com/books/2015/mar/13/john-gray-steven-pinker-wrong-violence-war-declining.

62 Howard Bloom, "Beyond the Supercomputer: Social Groups as Learning Machines," The Website of Howard Bloom, July 24, 2012, http://howardbloom.net/beyond-the-supercomputer-social-groups-as-learning-machines/.

63 Ibid.

64 Ibid.

65 Ibid.

66 Ibid.

67 Ibid.

68 Ibid.

69 April Holloway, "New Study Leaves Little Room for Doubt – Neanderthals and Humans Interbred," *Ancient Origins*, April 10, 2014, http://www.ancient-origins.net/news-evolution-human-origins/new-study-leaves-little-room-doubt-neanderthals-humans-interbred-088978.

[70] Melissa Hogenboom, "Human Evolution was Shaped by Interbreeding," *BBC*, October 13, 2015, Earth, http://www.bbc.com/earth/story/20151013-how-interbreeding-shaped-us.

## Chapter 7

[1] Mark Twain, *The Bible According to Mark Twain: Irreverent Writings on Eden, Heaven, and the Flood by America's Master Satirist*, eds. Joseph Mccullough and Howard Baetzhold (New York: Simon and Schuster, 1996), 76-77.

[2] Stephan Steinlechner, "Melatonin as a Chronobiotic: Pros and Cons," *Acta Neurobiologiae Experimentalis* 56 (1996): 363-372, http://www.ane.pl/pdf/56045.pdf.

[3] I. Y. Yoon, D. F. Kripke, J. A. Elliott, S. D. Youngstedt, "Luteinizing Hormone Following Light Exposure in Healthy Young men," *Neuroscience Letters* 341, no. 1 (2003): 25-28, doi: 10.1016/S0304-3940(03)00122-8.

[4] L. Bossini, et al., "Light Therapy as a Treatment for Sexual Dysfunctions," *Psychotherapy and Psychosomatics* 78, no. 2 (2009), doi: 10.1159/000203119.

[5] Konstantin V. Danilenko and Elena A. Samoilova, "Stimulatory Effect of Morning Bright Light on Reproductive Hormones and Ovulation: Results of a Controlled Crossover Trial," *PLoS Clinical Trials* 2, no. 2 (2007): e7, doi: 10.1371/journal.pctr.0020007.

[6] Padera Faryadyan, Afra Khosravi, Meysam Kashiri, and Reza Valizadeh, "Prenatal Exposure to Blue and Green Colors Increases Serum Testosterone Levels in Male Offspring Rats," *Biomedical & Pharmacy Journal* 7, no. 1 (2014): 129-135, http://www.biomedpharmajournal.org/dnload//BPJV07I01P129-135.pdf.

[7] J. Cao, et al., "Green and Blue Monochromatic Lights Promote Growth and Development of Broilers via Stimulating Testosterone Secretion and Myofiber Growth," *The Journal of Applied Poultry Research* 17, no. 2 (2008): 211-218, doi: /10.3382/japr.2007-00043.

[8] Jane E. Brody, "From Fertility to Mood, Sunlight Found to Affect Human Biology," *The New York Times*, June 23, 1981, http://www.nytimes.com/1981/06/23/science/from-fertility-to-mood-sunlight-found-to-affect-human-biology.html?pagewanted=all&mcubz=0.

[9] "Sunlight and Vitamin D improve mood," *DrDobbin Nutrition*, January 11, 2011, http://www.drdobbin.co.uk/sunlight-vitamin-d.

[10] Simon N. Young, "How to Increase Serotonin in the Human Brain without Drugs," *Journal of Psychiatry & Neuroscience* 32, no. 6 (2007): https://www.ncbi.nlm.nih.gov/pmc/articles/PMC2077351/.

[11] Maurice Cotterell, "How Our 28-day Spinning Sun Regulates Fertility in Females," The Website of Maurice Cotterell, last updated July 10, 2017, https://www.mauricecotterell.com/downloads/fertility.pdf.

[12] Matia B. Solomon, "Estrogen in the Brain Prevents Obesity and Glucose Intolerance during Menopause in Lab Animal Study," *Society for the Study of Ingestive Behavior*, July 18, 2017, http://www.ssib.org/web/press2017.php.

[13] Deborah Clegg, "Revealing Estrogen's Secret Role in Obesity," (paper presented at the national meeting of the American Chemical Society, Session "Genomics of Obesity." August 20, 2007), https://www.acs.org/content/acs/en/pressroom/newsreleases/2007/august/news-release-revealing-estrogens-secret-role-in-obesity.html.

[14] Ker Than, "Out-of-this-World Hypothesis: Cosmic Forces Control Life on Earth," *Space*, April 23, 2007, https://www.space.com/3721-world-hypothesis-cosmic-forces-control-life-earth.html.

[15] Ibid.

[16] Ibid.

[17] Michael Rampino, "The 30-Million-Year Mass Extinction Cycle – A Coincidence, or a Dark-Matter Event?" *The Daily Galaxy*, May 25, 2015, http://www.dailygalaxy.com/my_weblog/2015/05/the-30-million-year-mass-extinction-cycle-a-coincidence-or-a-dark-matter-event-holiday-feature-in-1980-walter-alvarez.html.

[18] D. Miller, J. Summers, and S. Silber, "Environmental Versus Genetic Sex Determination: A Possible Factor in Dinosaur Extinction?" *Fertility and Sterility* 81, no. 4 (2004): 954-964, doi: 10.1016/j.fertnstert.2003.09.051.

[19] Brian Switek, "Did Dinosaurs Die Out Because Males Couldn't Find a Date?" *Smithsonian*, February 18, 2011, https://www.smithsonianmag.com/science-nature/did-dinosaurs-die-out-because-males-couldnt-find-a-date-95200961/.

[20] Ibid.

[21] S. R. Thomas, M. J. Owens, M. Lockwood, "The 22-Year Hale Cycle in Cosmic Ray Flux - Evidence for Direct Heliospheric Modulation," *Solar Physics* 289, no. 1 (2014): 407-421, doi: 10.1007/s11207-013-0341-5.

[22] Ibid.

[23] Edward R. Dewey and Og Mandino, *Cycles: The Mysterious Forces That Trigger Events* (New York: Hawthorn Books, 1971), 57-59.

[24] Cotterell, "How Our 28-day Spinning Sun Regulates Fertility in Females," 4.

[25] Ibid.

[26] Alex Dominguez, "Egyptian Pharaoh Akhenaten Had Androgynous Build," *National Geographic News*, May 2, 2008, https://news.nationalgeographic.com/news/2008/05/080502-AP-feminine-ph.html.

[27] Simon Edge, "The Truth of King Tut: The Egyptian Pharaoh was Developing Breasts and had Wide Hips," *Express*, October 21, 2014, https://www.express.co.uk/news/history/525480/King-Tutankhamun-breasts-hips-club-foot-truth.

[28] *Wikipedia*, s.v. "Sunspot Numbers.png," last modified August 6, 2017, https://commons.wikimedia.org/wiki/File:Sunspot_Numbers.png.

[29] J. A. Eddy, "The Maunder Minimum," *Science* 192, no. 4245 (1976): 1189-1202, doi: 10.1126/science.192.4245.1189.

[30] "Jonathan Cahn Unlocks Thousands of Years of Mystery," *CBN*, http://www1.cbn.com/700club/jonathan-cahn-unlocks-thousands-years-mystery.

[31] Natalie Wolchover, "Did Cold Weather Cause the Salem Witch Trials?" *LiveScience*, April 20, 2012, https://www.livescience.com/19820-salem-witch-trials.html.

[32] "Yearly Mean and Monthly Smoothed Sunspot Number," *Sunspot Index and Long-term Solar Observations*, Royal Observatory of Belgium, Brussels, last modified February 1, 2018, http://www.sidc.be/silso/yearlyssnplot.

## Chapter 8

[1] Nell Greenfieldboyce, "Did Climate Inspire the Birth of a Monster?" *NPR*, August 13, 2007, https://www.npr.org/2007/08/13/12688403/did-climate-inspire-the-birth-of-a-monster.

[2] David Archibald, "The Potential Impact of Volcanic Overprinting of the Eddy Minimum," *Watts Up with That?* (blog), January 7, 2013, https://wattsupwiththat.com/2013/01/07/the-potential-impact-of-volcanic-overprinting-of-the-eddy-minimum/.

[3] Ibid.

[4] Henrik Svensmark, "Cosmoclimatology: A New Theory Emerges," *Astronomy & Geophysics* 48, no. 1 (2007): 1.18-1.24, doi: 10.1111/j.1468-4004.2007.48118.x.

[5] Petra Breitenmoser, et al., "Solar and Volcanic Fingerprints in Tree-Ring Chronologies over the Past 2000 Years," *Palaeogeography, Palaeoclimatology, Palaeoecology* 313-314 (2012): 127-139, doi: 10.1016/j.palaeo.2011.10.014.

[6] Ibid.

[7] Sacha Dobler, "Black Death and Abrupt Earth Changes in the 14th Century," *Abrupt Earth Changes*, May 25, 2017, https://abruptearthchanges.files.wordpress.com/2017/05/black-death-and-abrupt-earth-changes-in-the-14th-century-pdf.pdf.

[8] Ibid.

9 Tony Phillips, "Deep Solar Minimum," *NASA Science*, April 1, 2009, https://science.nasa.gov/science-news/science-at-nasa/2009/01apr_deepsolarminimum.

10 Tony Phillips, "Solar Wind Loses Power, Hits 50-year Low," *NASA Science*, September 23, 2008, https://science.nasa.gov/science-news/science-at-nasa/2008/23sep_solarwind.

11 Ibid.

12 Peter Wehrwein, "Astounding Increase in Antidepressant Use by Americans," *Harvard Health* (blog), October 20, 2011, https://www.health.harvard.edu/blog/astounding-increase-in-antidepressant-use-by-americans-201110203624.

13 Robert Prechter, *The Socionomic Theory of Finance* (Gainesville, GA: Socionomics Institute Press 2016), 175, 182.

14 Marjolein van Baardwijk and Philip Hans Franses, "The Hemline and the Economy: Is There Any Match?" *Erasmus School of Economics*, July 28, 2010, hdl.handle.net/1765/20147.

15 Prechter, The Socionomic Theory of Finance, 182.

16 Robert R. Prechter, Jr. and Peter Kendall, "The Sunspot Cycle and Stocks," *Eliott Wave Theorist*, September 2000, http://www.socionomics.net/2011/09/article-the-sunspot-cycle-and-stocks/.

17 *Wikimedia Commons*, s.v. "Hathaway Cycle 24 Prediction," last modified September 3, 2014, https://commons.wikimedia.org/wiki/File:Hathaway_Cycle_24_Prediction.png

18 Simon J. Shepherd, Sergei I. Zharkov, and Valentina V. Zharkova, "Prediction of Solar Activity from Solar Background Magnetic Field Variations in Cycles 21-23," *The Astrophysical Journal* 795, no. 1 (2014): doi: 10.1088/0004-637X/795/1/46.

19 John Coates, *The Hour Between Dog and Wolf: Risk Taking, Gut Feelings and the Biology of Boom and Bust* (London: Penguin Press, 2012), 27-28.

20 Ibid., 28.

21 Ibid.

22 Daniella Silva, "'Fearless Girl' Statue will Face off Wall Street Bull for Another Year," *NBC News*, March 27, 2017, http://www.nbcnews.com/news/us-news/fearless-girl-statue-will-face-wall-street-bull-another-year-n738991.

23 Hazel Torres, "'The Snapping of the American Mind': New Book says Obama is Brainwashing America," *Christianity Today*, October 17, 2015, Society, https://www.christiantoday.com/article/the-snapping-of-the-american-mind-new-book-says-obama-is-brainwashing-america/67843.htm.

24 N.D. Volkow, et al., "Imaging Dopamine's Role in Drug Abuse and Addiction," *Neuropharmacology* 56, supplement 1 (2009): 3-8, doi: 10.1016/j.neuropharm.2008.05.022.

25 Ibid.

26 Dennis Baron, "The Gender-Neutral Pronoun: 150 Years Later, Still an Epic Fail," *OUP Blog*, August 26, 2010, https://blog.oup.com/2010/08/gender-neutral-pronoun.

27 "We Added a Gender-Neutral Pronoun in 1934. Why Have so Few People Heard of It?" *Merriam Webster*, https://www.merriam-webster.com/words-at-play/third-person-gender-neutral-pronoun-thon.

28 Astead W. Herndon, "Harvard Allows Students to Pick New Gender Pronouns," *Boston Globe*, September 2, 2015, Metro, https://www.bostonglobe.com/metro/2015/09/02/harvard-allows-students-pick-new-gender-pronouns/C0EXpZHw09zwCz04hVhjdJ/story.html.

29 Eugene Volokh, "You Can be Fined for not Calling People 'Ze' or 'Hir,' if That's the Pronoun They Demand that You Use," *Washington Post*, May 17, 2016, Opinions https://www.washingtonpost.com/news/volokh-conspiracy/wp/2016/05/17/you-can-be-fined-for-not-calling-people-ze-or-hir-if-thats-the-pronoun-they-demand-that-you-use/.

# Chapter 9

1 Paglia, *Sexual Personae*, 96.

2 Stephanie Pappas, "Brain Sees Men as Whole, Women as Parts," *Live Science*, July 24, 2012, http://www.livescience.com/21806-brain-male-female-objectification.html.

3 Martha C. Nussbaum, "Objectification," *Philosophy and Public Affairs* 24, no. 4 (1995): 227, http://www.mit.edu/~shaslang/mprg/nussbaumO.pdf.

4 Sandra Harding, *The Science Question in Feminism* (Ithaca, NY: Cornell University Press, 1986), 113.

5 Peter Harrison, "A God of Math & Order," *Christianity Today*, 2002, Christian History, http://www.christianitytoday.com/history/issues/issue-76/god-of-math-order.html.

6 Isaac Newton, "Book 3: The System of the World; General Scholium," *Philosophical Writings*, Andrew Janiak ed. (Cambridge, UK: Cambridge University Press, 2004): Section XI.

7 Ibid.

8 Marcelo Dascal and Victor D. Boantza, eds., *Controversies within the Scientific Revolution* (Amsterdam: John Benjamins Publishing, 2011), 108.

9 Ibid.

10 *Wikipedia*, s.v. "Yin and Yang," last modified January 17, 2018, https://en.wikipedia.org/wiki/Yin_and_yang.

11 Ludwig Wittgenstein, *Culture and Value*, trans. Peter Winch (Chicago: University of Chicago Press, 1984), 18c.

# Sex Wars

[12] Rudolf Steiner, "The Phenomena of the World of Colors," in *Goethe's World View*, trans. William Lindeman (Spring Valley, NY: Mercury Press, 1985), http://wn.rsarchive.org/Books/GA006/English/APC1928/GA006_c08.html.

[13] Stephen J. Puetz and Glenn Borchardt, *Universal Cycle Theory: Neomechanics of the Hierarchically Infinite Universe* (Denver, CO: Outskirts Press, 2011).

[14] *Wikipedia*, s.v. "The Tao of Physics," last modified November 12, 2017, https://en.wikipedia.org/wiki/The_Tao_of_Physics.

[15] Amara Graps, "What is the Speed of the Solar System?" *Stanford University*, http://solar-center.stanford.edu/FAQ/Qsolsysspeed.html.

[16] DjSadhu, "The Helical Model – Vortex Solar System Animation" [video], The Website of DjSadhu, last updated January 23, 2018,http://www.djsadhu.com/the-helical-model-vortex-solar-system-animation/.

[17] Benjamin Franklin, "Aurora Borealis: Suppositions and Conjectures towards Forming an Hypothesis for its Explanation," in *Memoirs of the life and Writings of Benjamin Franklin* (London, Henry Colburn, 1818): 291-297.

[18] Hannes Alfven, "Double Layers and Circuits in Astrophysics," *Transactions on Plasma Science* 14, no. 6 (1986): 779-793, doi: 10.1109/TPS.1986.4316626.

[19] Albert Einstein, "Ether and Relativity," in *Sidelights on Relativity* (London: Methuen & Co, 1922).

[20] Edmund Taylor Whittaker, *A History of the Theories of Aether and Electricity from the Age of Descartes to the Close of the 19th Century* (London: Longmans, Green and Company, 1910), 101-102.

[21] Gerald H. Pollack, *The Fourth Phase of Water: Beyond Solid, Liquid, and Vapor* (Seattle: Ebner and Sons, 2013).

[22] Ibid., xviii.

[23] Ibid.

[24] Ibid., 3.

[25] Ibid., 333.

[26] Ibid.

[27] Ibid., 335.

[28] Eric J. Chaisson, "The Natural Science Underlying Big History," *Scientific World Journal* 2014, no. 1 (2014): 1, doi: 10.1155/2014/384912.

[29] Ibid.

[30] Ibid.

[31] Ibid.

[32] Eric J. Chaisson, "Energy Rate Density as a Complexity Metric and Evolutionary Driver," *Complexity* 16, no. 3 (2010): 27-40, doi: 10.1002/cplx.20323.

[33] *Wikipedia*, s.v. "Kardashev Scale," last modified January 22, 2018, https://en.wikipedia.org/wiki/Kardashev_scale.

[34] *Wikipedia*, s.v. "Second Law of Thermodynamics," last modified January 14, 2018, https://en.wikipedia.org/wiki/Second_law_of_thermodynamics.

[35] William Thomson [1st Lord Kelvin], "On the Dynamical Theory of Heat, with Numerical Results Deduced from Mr. Joule's Equivalent of a Thermal Unit, and M. Regnault's Observations on Steam," *London and Edinburgh Philosophical Magazine and Journal of Science* IV, no. 22 (1852): 13, https://archive.org/stream/londonedinburghp04maga#page/12/mode/2up.

[36] Rudolf Clausius, "On a Modified Form of the Second Fundamental Theorem in the Mechanical Theory of Heat," trans. in *London, Edinburgh, and Dublin Philosophical Magazine and Journal of Science* 4, no. 2 (1856):86, https://www.biodiversitylibrary.org/item/20044#page/100/mode/1up.

[37] Gregory Hill and Kerry Thornley, *Principia Discordia: How I Found Goddess and what I Did to Her when I Found Her* (New Orleans: 1965; reprint, London: Forgotten Books, 2007), 67.

[38] *Wikipedia*, s.v. "Photoelectric Effect," last modified January 21, 2018, https://en.wikipedia.org/wiki/Photoelectric_effect.

[39] A. Wayne Meikle, "The Interrelationships between Thyroid Dysfunction and Hypogonadism in Men and Boys," *Thyroid* 14, supplement 1 (2004): 17-25, doi: 10.1089/105072504323024552.

# Chapter 10

[1] Karl R. Popper, *The Logic of Scientific Discovery* (Toronto: University of Toronto Press: 1959), 15.

[2] Eric D. Beinhocker, "Reflexivity, Complexity, and the Nature of Social Science," *Journal of Economic Methodology* 20, no. 4 (2013): 330-342, doi: 10.1080/1350178X.2013.859403.

[3] David S. Wilson, *Darwin's Cathedral: Evolution, Religion, and the Nature of Society* (Chicago: University Of Chicago Press, 2003).

[4] Czarina Ong, "Richard Dawkins Says Christianity is World's Best Defense against Radical Islam," *Christianity Today*, January 13, 2016, https://www.christiantoday.com/article/richard.dawkins.says.christianity.is.worlds.best.defence.against.radical.islam/76416.htm.

[5] Jacey Fortin, "Richard Dawkins Event Canceled over Past Comments about Islam," *New York Times*, July 24, 2017, U.S. edition, https://www.nytimes.com/2017/07/24/us/richard-dawkins-speech-canceled-berkeley.html.

[6] Jonathan Haidt, "Moral Psychology and the Law: How Institutions Drive Reasoning, Judgement, and the Search for Evidence," *Alabama Law Review* 64, no. 4 (2013): 867-880,

https://www.law.ua.edu/pubs/lrarticles/Volume%2064/Issue%204/4%20Haid
t%20867-880.pdf.

[7] *Wikipedia*, s.v, "Cult of Reason," last modified February 1, 2014,
http://en.wikipedia.org/wiki/Cult_of_Reason#cite_ref-Carlyle379_11-0.

[8] Roy Barzilai, *The Objective Bible: Western Civilization's Struggle for Philosophic
Liberation from a Herd-Mentality and Pagan Mysticism* (Dibrah Publishing,
2014), 78.

[9] Jeffrey A. Tucker, "The Prehistory of the Alt-Right," *Foundation for Economic
Education*, March 8, 2017, https://fee.org/articles/the-prehistory-of-the-alt-
right/.

[10] *Wikipedia*, s.v. "Extinction," last modified January 2, 2018,
https://en.wikipedia.org/wiki/Extinction.

[11] Tucker, "The Prehistory of the Alt-right."

[12] Stephanie Pain, "Blasts from the Past: The Soviet Ape-Man Scandal," *New
Scientist*, August 20 2008, Histories,
https://www.newscientist.com/article/mg19926701-000-blasts-from-the-past-
the-soviet-ape-man-scandal/.

[13] David Levy, *Love and Sex with Robots: The Evolution of Human-Robot
Relationships* (New York, Harper Perennial, 2008).

[14] Jon Haidt, Lee Jussim, and Chris Martin, "The Problem," *Heterodox Academy*,
September 8, 2015, http://heterodoxacademy.org/problems/.

[15] Edward Szall, "Futurist Ray Kurzweil: AI-Human Merger only 12 Yrs Away,"
*TruNews*, March 16, 2017, https://www.trunews.com/article/futurist-ray-
kurzweil-ai-human-merger-only-12-yrs-away.

[16] Gary Marcus, "Artificial Intelligence is Stuck. Here's How to Move it Forward,"
*New York Times*, July 29, 2017, Opinion,
https://www.nytimes.com/2017/07/29/opinion/sunday/artificial-intelligence-is-
stuck-heres-how-to-move-it-forward.html.

[17] *Wikipedia*, s.v. "AI Winter," last modified January 10, 2018,
https://en.wikipedia.org/wiki/AI_winter.

[18] Shivali Best, "Tesla's Elon Musk Warns We Only have 'a 5 to 10% Chance' of
Preventing Killers Robots from Destroying Humanity," *Daily Mail*, November
23, 2017, Science, http://www.dailymail.co.uk/sciencetech/article-5110787/Elon-
Musk-says-10-chance-making-AI-safe.html.

[19] Alan Winfield, "Artificial Intelligence will not Turn into a Frankenstein's
Monster," *The Guardian*, August 9, 2014, Tech,
https://www.theguardian.com/technology/2014/aug/10/artificial-intelligence-
will-not-become-a-frankensteins-monster-ian-winfield.

[20] Samuel Gibbs, "Apple Co-Founder Steve Wozniak Says Humans will be
Robots' Pets," *The Guardian*, June 25, 2015, Tech,

https://www.theguardian.com/technology/2015/jun/25/apple-co-founder-steve-wozniak-says-humans-will-be-robots-pets.

[21] Howard Bloom, *The God Problem: How a Godless Cosmos Creates* (Amherst, NY: Prometheus Books, 2012).

Printed in Great Britain
by Amazon

56256396R00175